RIPPLES IN SPACETIME

RIPPLES IN SPACETIME

$(((\quad)))$

Einstein, Gravitational Waves, and
the Future of Astronomy

GOVERT SCHILLING

The Belknap Press of Harvard University Press

Cambridge, Massachusetts | London, England

2017

Library of Congress Cataloging-in-Publication Data

Names: Schilling, Govert, author.
Title: Ripples in spacetime : Einstein, gravitational waves, and the future
   of astronomy / Govert Schilling.
Description: Cambridge, Massachusetts : The Belknap Press of Harvard
   University Press, 2017. | Includes bibliographic references and index.
Identifiers: LCCN 2017007571 | ISBN 9780674971660 (alk. paper)
Subjects: LCSH: Gravitational waves. | Relativistic
   astrophysics—Methodology. | Einstein, Albert, 1879–1955.
Classification: LCC QC179 .S287 2017 | DDC 539.7/54—dc23
LC record available at https://lccn.loc.gov/2017007571

Contents

Foreword

MARTIN REES

Einstein has a unique place in the pantheon of science—and deservedly so. His insights into space and time have transformed our understanding of gravity and the cosmos. Everyone recognizes the benign and unkempt sage of poster and T-shirt. But his best work was done when he was young. He was still in his thirties when he was catapulted to worldwide fame. On May 29, 1919, there was a solar eclipse. A group led by the astronomer Arthur Eddington observed stars appearing close to the Sun during the eclipse. The measurements showed that these stars were displaced from their normal positions, the light from them being bent by the Sun's gravity. This confirmed one of Einstein's key predictions. When these results were reported at the Royal Society in London, the world press spread the news. "Lights All Askew in the Heavens; Einstein Theory Triumphs" was the rather over-the-top headline in the *New York Times*.

Einstein's theory of general relativity, put forward in 1915, is a triumph of pure thought and insight. Its implications for us on Earth are slight. It requires minor adjustments in the clocks used in modern navigation systems, but Newton remains good enough for launching and tracking space probes.

In contrast, Einstein's insight that space and time are linked—that "space tells matter how to move; matter tells space how to curve"—is crucial to many cosmic phenomena. But it's hard to test a theory whose consequences are so remote. For almost half a century after it was proposed, general relativity was sidelined from the mainstream of physics. But from the 1960s onward, evidence has grown for a "big bang" that set the universe expanding and for black holes—two of Einstein's key predictions.

And in February 2016, nearly a hundred years after the famous Royal Society meeting reporting the eclipse expedition, another announcement—this time at the Press Club in Washington, DC— further vindicated Einstein's theory. This was the detection of gravitational waves by LIGO (Laser Interferometer Gravitational-Wave Observatory). This is the theme of Govert Schilling's book. He has a wonderful story to tell, spanning more than a century.

Einstein envisaged the force of gravity as a "warping" of space. When gravitating objects change their shapes, they generate ripples in space itself. When such a ripple passes the Earth, our local space "jitters": it is alternately stretched and compressed as gravitational waves pass through it. But the effect is tiny. This is because gravity is such a weak force. The gravitational pull between everyday objects is minuscule. If you wave around two dumbbells, you will emit gravitational waves—but with quite infinitesimal power. Even planets orbiting stars, or pairs of stars orbiting each other, don't emit at a detectable level.

Astronomers are agreed that the only sources that LIGO might detect must involve much stronger gravity than in ordinary stars and planets. The best bet is that the events involve black holes. We've known for nearly fifty years that black holes exist: most are the remnants of stars twenty or more times more massive than the Sun. These stars burn brightly, and in their explosive death throes (signaled by a supernova), their inner part collapses to a black hole. The material that the star was made of gets cut off from the rest of the universe, leaving a gravitational imprint on the space it's left.

If two black holes were to form a binary system, they would gradually spiral together. As they get closer, the space around them becomes more distorted until they coalesce into a single, spinning hole.

This hole sloshes and "rings," generating further waves until it settles down as a single, quiescent hole. It is this "chirp"—a shaking of space that speeds up and strengthens until the merger, and then dies away— that LIGO can detect. These cataclysms happen less than once in a million years in our galaxy. But such an event would give a detectable LIGO signal even if it happened a billion light-years away—and there are millions of galaxies closer than that. To detect even the most propitious events requires amazingly sensitive—and very expensive— instruments. In the LIGO detectors, intense laser beams are projected along four-kilometer-long vacuum pipes and reflected from mirrors at each end. By analyzing the light beams, it's possible to detect changes in the distance between the mirrors, which alternately increases and decreases as space expands and contracts. The amplitude of this vibration is exceedingly small, about 0.0000000000001 centimeters—millions of times smaller than the size of a single atom. The LIGO project involves two similar detectors about 3,000 kilometers apart—one in Washington State, the other in Louisiana. A single detector would register microseismic events, passing vehicles, et cetera. To exclude these false alarms, experimenters take note only of events that show up in both.

For years, LIGO detected nothing. But it went through an upgrade, coming fully on line again in September 2015. After literally decades of frustration, the quest succeeded: a chirp was detected that signaled the collision of two black holes more than a billion light-years away, and it opened up a new field of science—probing the dynamics of space itself.

It is sadly not unknown for hyped-up scientific claims to be mistaken or exaggerated—and the book recounts such claims in this field, too. I count myself a hard-to-convince skeptic. But what the LIGO researchers claimed—the culmination of literally decades of effort by scientists and engineers with high credentials—is compelling, and this time I expect to be fully convinced.

This detection is indeed a big deal. It's one of the great discoveries of the decade, up there with the detection of the Higgs particle, which caused huge razzmatazz in 2012. The Higgs particle was a capstone to the Standard Model of particle physics, developed over several decades. Likewise, gravitational waves—vibrations in the fabric of space itself—are a crucial and distinctive consequence of Einstein's theory of general relativity.

Peter Higgs predicted his particle fifty years ago, but its detection—and the pinning down of its properties—had to await the march of technology. It required a huge machine, the Large Hadron Collider in Geneva. Gravitational waves were predicted even earlier, but detection has been delayed because the quest involves detecting a very elusive effect and again requires large-scale and ultra-precise equipment.

Quite apart from offering a completely new vindication of Einstein's theory, these results deepen our understanding of stars and galaxies. Astronomical evidence on black holes and massive stars is limited—it was hard to predict how many would be within range. Pessimists thought that the events might be so rare that even the new and improved LIGO wouldn't detect anything, at least for a year or two. But unless the experimenters have had exceptional "beginners' luck," it looks as though a new kind of astronomy has been founded, revealing the dynamics of space itself rather than the material that pervades it. Other detectors in Europe, India, and Japan have been included in the search, and there are plans to launch detectors into space.

But all too many scientists shy away from trying to explain their ideas and discoveries, believing that they're arcane and incomprehensible. It's true that professional scientists express their ideas via mathematics—a foreign language to many people. But the key ideas can be expressed in simple language by sufficiently skilled writers. Govert Schilling is one of the best of these, and he has surpassed himself in this book. His narrative spans more than a century. The book explains the key concepts in clear and entertaining terms, setting them in historical context. He also depicts the diverse personalities involved. Some have been "obsessives," but this is unsurprising—indeed obsessiveness is a prerequisite for anyone devoting years, even decades, to an experimental challenge with no guarantee of any payoff. But the effort has been supported by hundreds of experts cooperating in teams. He tells of boisterous controversies, setbacks, and amazing technical achievements by scientists and engineers who struggled for decades to achieve fantastic precision. He relates how they emerged triumphant, revealing clues to the bedrock nature of space and time. It is a wonderful story, engagingly and rivetingly told.

Introduction

Orbiting a yellow dwarf star on the outskirts of a spiral galaxy is a tiny planet formed some 3.3 billion years ago from accumulating dust and pebbles. Organic compounds that rained down from outer space into the blue planet's lukewarm oceans have arranged themselves into self-replicating molecules. By now, those waters are brimming with single-celled life-forms. It won't be too long before life finds its way onto the planet's barren continents.

In another corner of this vast universe, two extremely massive stars have ended their brief lives in catastrophic supernova explosions. What's left is a tight binary of voracious black holes, each tens of times more massive than the faraway yellow dwarf star. Their gravity pulls in gas and dust that venture too close, and bends the paths of light rays in their vicinity. Nothing will ever be able to escape the tight gravitational grip of these cosmic abysses.

As the black holes orbit each other, they're making waves: minute ripples in spacetime that propagate with the speed of light. The waves carry away energy, causing the two holes to draw closer and closer. Eventually they orbit each other hundreds of times per second at half the speed of light. Spacetime is stretched and squeezed; the tiny perturbations grow into massive waves. Then, in a final burst of pure energy, the two black holes collide and merge into one. Quiet returns to the scene. But the last powerful waves are spreading into space like a tsunami.

It takes the death cry of the two black holes 1.3 billion years to reach the outskirts of our spiral galaxy. By then, their amplitude has diminished tremendously. They still push and pull on everything that's in their way, but nobody would ever notice. On the blue planet, ferns and trees now cover the surface; an asteroid impact has wiped out giant reptiles, and one of the many lines of mammals living on this world has evolved into a species of curious two-legged creatures.

Having passed the outer regions of the Milky Way galaxy, the gravitational waves from the distant black hole merger take only 100,000 years or so to reach the vicinity of the Sun and the Earth. As they race toward the planet at 300,000 kilometers per second, people start to explore the universe they're part of. They grind telescope lenses, discover new planets and moons, and map the Milky Way.

One hundred years before the waves arrive—they've covered 99.99999 percent of their 1.3-billion-light-year trip—a twenty-six-year-old scientist named Albert Einstein predicts their potential existence. It takes another half century before people seriously start to hunt them down. Finally, in the early twenty-first century, detectors have become sensitive enough. Just a few days after being switched on, they register the tiny vibrations, much smaller in amplitude than the size of an atomic nucleus.

On Monday, September 14, 2015, at 09:50:45 Universal Time, a hundred-year-old prediction of Einstein is borne out as astronomers secure a gravitational message from a black hole collision in a galaxy far, far away.

The first direct detection of a gravitational wave is rightly hailed as one of the greatest scientific discoveries of the new century. Further detections, with ever more sensitive equipment, will provide astronomers with a completely new way of studying the violent universe and offer physicists the opportunity to finally solve the secrets of spacetime.

A few years before the latest version of the Laser Interferometer Gravitational-Wave Observatory (LIGO) went online, I first thought about writing this book. Wouldn't it be great, I thought, if the first observation of a gravitational wave took place right around the time

I was finishing my manuscript? The book could then be published shortly after the announcement, with an added epilogue on the new result.

Science progressed faster than I had expected. Almost no one could have imagined that the new detector would win the grand prize during its very first days of operation. So most of my research and all of my writing had to take place *after* the monumental find. Now that the book is finished, I'm happy about the timing—the discovery has become an integral part of the story, as opposed to an afterthought.

The history of gravitational wave astronomy has been told before. In this book, however, it's just half of the story. *Ripples in Spacetime* is very much about science in progress, about the way discoveries are made, about events that take place today, and about expectations for the future, when studying gravitational waves will have become a mature field of astronomy. The discovery of GW150914—the signal that was picked up on that memorable Monday—is both the culmination of a century-long quest and the starting point for a completely new chapter in our exploration of the universe.

(((1)))

A Spacetime Appetizer

Joe Cooper dons his NASA spacesuit and puts on his helmet. He'll need oxygen available in case something goes wrong during the launch. Technicians help him step into the spacecraft, perched atop the towering rocket. Through his radio he hears the countdown sequence, and he feels the adrenaline flow through his veins. Cooper is no wimp, but being shot into space on top of a pillar of flame is always a bit unnerving.

Soon enough he and his three fellow astronauts are on their way. Everything goes smoothly. Outside the small windows of their vehicle, blue sky gives way to black void. The engines shut down; weightlessness sets in. Now all they have to do is catch up with the huge spaceship that's circling Earth at more than 8 kilometers per second, and then dock. Easy.

This all sounds like a regular trip to the International Space Station (ISS) on board a Russian Soyuz craft. Business as usual . . . or is it? You've never heard of a NASA astronaut named Joe Cooper. And Cooper can't have three crewmates. As every astronaut can tell you, the Soyuz is much too small to hold four people—even with three it's crowded.

Then you hear the next part of this tale: The spaceship they dock with is called *Endurance,* and it doesn't look like the ISS at all. Finally, the astronauts fly *Endurance* to Saturn, disappear through a wormhole, emerge in another galaxy, orbit a giant black hole called Gargantua, and visit alien planets. Cooper even dives into hyperspace. Clearly, something's up.

This scenario is taken from the 2014 Hollywood blockbuster *Interstellar*, directed by Christopher Nolan, and astronaut Cooper is played by actor Matthew McConaughey. If you're a real space nerd, you may have recognized the name Joe Cooper. You may even have seen *Interstellar* more often than I have. It's a great movie.

One of the things that sets *Interstellar* apart from other sci-fi movies is its lineup of executive producers. There's Jordan Goldberg (*Batman, Inception*), Jake Myers (*The Revenant*), and Thomas Tull (*Jurassic World*). And then there's Kip Thorne, emeritus Feynman Professor of Theoretical Physics at the California Institute of Technology in Pasadena. Not many theoretical physicists moonlight as film producers.

What happens when a scientist gets involved with producing a science fiction movie? You'd hope the movie would get the science right. And it does, to an impressive degree. Thorne helped create the story line. He briefed the screenwriter, the director, the visual effects team, and the actors on astronomy and general relativity. He even wrote out the equations on the blackboard of movie professor John Brand (played by Michael Caine). Sadly, Thorne doesn't have a cameo appearance in the movie. Then again, KIPP, one of the robots, is evidently named after him.

Few people would be better equipped to be the science adviser for a movie about black holes than Kip Thorne. If anyone understands the bizarre properties of spacetime, it's he. In 1990, he even won a fifteen-year-old wager with his British colleague and friend Stephen Hawking about the true nature of an astronomical X-ray source known as Cygnus X-1. (The prize: a year-long subscription to *Penthouse* magazine.) Thorne's 1994 book *Black Holes and Time Warps* became a national bestseller.

In early 2016, Thorne's name was all over the place again. On February 11, scientists announced the first successful direct detection of gravitational waves. In the distant universe, two black holes had collided and merged. The crash sent ripples through spacetime. After a journey across more than a billion light-years, these waves arrived on Earth on September 14, 2015. The two giant Laser Interferometer Gravitational-Wave Observatory (LIGO) detectors in the United States recorded the infinitesimal quivering. And LIGO is the brainchild of Thorne and his physicist colleagues Rainer Weiss and Ronald Drever.

No one has ever seen a black hole up close. No one knows if wormholes really exist. Gravitational waves are much too faint to be detected without incredibly sensitive instruments. The bending of space, the slowing of time—it's all way too complicated and way beyond our daily experience. To really understand these things, you need to master Albert Einstein's theory of general relativity.

There's a famous anecdote about English astronomer Arthur Stanley Eddington. In the early twentieth century, Eddington was one of the biggest popularizers of Einstein's new theory of spacetime—we'll meet him again in Chapter 3. After a public lecture, a person in the audience asked him, "Professor Eddington, is it true that there are only three people in the world who really understand general relativity?" Eddington thought for a while, then replied, "Who might possibly be the third?"

Well, it's not *that* hard, of course. Tens of thousands of theoretical physicists all over the world understand the fundamentals of general relativity. Then again, new theoretical insights keep popping up all the time, especially in the realm of black holes, where quantum effects become important. There's Stephen Hawking's theory of evaporating black holes, Kip Thorne's wormhole shortcuts, Gerard 't Hooft's holographic principle, and Leonard Susskind's firewall.

I'm not going into the details here, but if the greatest minds alive keep finding new and surprising insights (and keep arguing about them), it's clear that the full scope of general relativity is still beyond their grasp. The examples given here are just a few of the less farfetched ideas. The journal *Physical Review Letters* also runs papers on eleven-dimensional spacetime, time travel, and the multiverse. And you thought *Interstellar* was speculative?

Maybe this is why so many people are interested in this stuff, even though it might appear to be rather useless knowledge. You don't need to know about black holes to run for president, and gravitational waves are not going to solve the problem of global warming. We can live out our lives without a care for general relativity. (There's one striking exception, but I'll save that for Chapter 3.) Still, it's exciting and intriguing, and it surely fires the imagination. Maybe that's reason enough.

Moreover, general relativity tells us about the way the world works at its most fundamental level. And isn't the quest to really understand our world one of the things that sets us apart from other animals?

To be honest, we weren't very good at understanding our world for many thousands of years. The first agricultural societies arose some 12,000 years ago in the Middle East. Back then, people were well aware of the cyclic motions of the Sun and the Moon. They saw patterns in the stars. They even noticed a handful of bright stars slowly wandering across the constellations. But that was about it. They had no clue as to the true nature of the celestial bodies. No urge to know, even. Sun, Moon, and planets were seen as gods—above and beyond our everyday world.

Nothing much changed until the time of the great Greek philosophers, some 2,500 years ago. That's 9,500 years—many hundreds of generations—without significant progress. If we compress 12,000 years of history into a single twenty-four-hour day, starting at midnight, we're already past 7:00 PM by the time Aristotle proposes the first model of the universe, consisting of nested crystal spheres. Our ancestors had the brainpower—after all, they were *Homo sapiens* just like us. They just didn't care enough.

The Greeks *did* care. They correctly inferred that the Earth is a sphere. They even determined its circumference with surprising accuracy. (Some schoolbooks still would have you believe that Christopher Columbus was the first person to discover the true shape of the Earth, but that's plain wrong.) And even though the Greeks didn't know what the Sun, the Moon, the planets, or the stars *were,* at least they tried to understand their intricate motions.

This culminated in the geocentric worldview of Claudius Ptolemy, who lived some nineteen centuries ago in what is now northern Egypt. (On our twenty-four-hour timeline since the rise of agriculture, that's around 8:10 PM.) As the name implies, Ptolemy's model had the Earth in the center. The Sun, Moon, and planets circled around the Earth in a complicated assembly of primary and secondary orbits. Ptolemy's worldview even explained why the planets appeared to be moving backward every now and then.

Nice try, but wrong. It took ages before people realized that something was amiss. That didn't happen until Polish astronomer Nicolaus Copernicus published his heliocentric worldview, centered on the Sun instead of the Earth, in 1543—just past 11:00 PM on our condensed timeline. Understanding our world was a frustratingly slow process for most of the past 12,000 years.

Soon after Copernicus, however, the rate accelerated. Scientists discovered that the book of nature is written in the language of mathematics, as Italian physicist Galileo Galilei so nicely put it. Galileo studied how stuff moves: he proved Aristotle wrong on a number of his assumptions, and he used mathematical equations to describe his findings. Not much later, in Germany, Johannes Kepler formulated his famous laws of planetary motion.

What does this history have to do with black holes, gravitational waves, and the secrets of spacetime? Everything. Copernicus, Galileo, and Kepler laid the groundwork for Isaac Newton's theory of universal gravitation, first published in 1687. And Albert Einstein's general relativity—the theory behind *Interstellar*—replaced the old ideas of Newton. Our understanding the world is made possible only by improving on the work of others. The crystal spheres of Aristotle and the wormholes of Kip Thorne are connected by one great arc of clever thinking and discovery.

Another revolution occurred in the early seventeenth century—a revolution of tools. Dutch spectacle maker Hans Lipperhey invented the telescope, but the use of the new instrument was subsequently pioneered by Galileo. Galileo discovered craters and mountains on the Moon, dark spots on the Sun, satellites orbiting Jupiter, and countless stars in the Milky Way. Eventually, ever-larger telescopes brought us binary stars, asteroids, nebulae, and galaxies—and black holes, of course. Without the telescope, astronomy would still be in its infancy.

Let's do a quick virtual tour of the cosmos to make sure we have the general picture right.

Earth is a planet. Like seven other planets, it orbits the Sun. The four innermost planets (Mercury, Venus, Earth, and Mars) are really small; they consist of metals and rocks. The outer four (Jupiter, Saturn, Uranus, and Neptune) are much bigger; they mainly consist of gas and ice. Between the orbits of Mars and Jupiter is a belt of asteroids—rocky leftovers from the birth of the solar system. Beyond Neptune there's another debris belt of ice balls and frozen dwarf planets, of which Pluto is the largest.

Look up during the daytime, and you'll see a huge sphere of incandescent gas in the sky—the Sun. The planets in the solar system receive all their light and heat from the Sun. Look up at night, and you'll see thousands of other suns—the stars. They appear small, faint, and cold, but that's just because they are tremendously far away. Put the Sun at a similar distance, and all you'd see is yet another small pinprick of light.

In Chapter 5, I'll tell you much more about stars. For now, remember that every star is a sun and that most of them are likely to be accompanied by their own family of planets. In fact, at the time of writing, well over 3,000 exoplanets have been discovered.

Too bad we can't go there and take a look, at least not in the foreseeable future. Even light, traveling at 300,000 kilometers per second, would take 4.3 years to travel from the Sun to the nearest star, Proxima Centauri. That's why astronomers say that Proxima is at a distance of 4.3 light-years. (One light-year is $300,000 \times 60 \times 60 \times 24 \times 365.25$ kilometers. That's almost 9.5 trillion kilometers.)

Have you ever tried to count the stars in the night sky? With the naked eye you can see a few thousand of them, depending on how dark your sky is. Most of them are a few tens or a few hundreds of light-years away, which is incredibly distant for most human beings but relatively close for astronomers—our cosmic backyard.

The vast majority of stars in our Milky Way galaxy are much farther away. To see them, you'll have to use a telescope. They come in many colors and sizes, and their names—red dwarfs, white dwarfs, yellow subgiants, blue supergiants—evoke the inhabitants of a fairy forest. And there are many of them. Astronomers now think that the Milky Way contains several hundred billion stars. One of them is our Sun.

We're not done yet. Our Milky Way galaxy is not alone. The universe is filled with other galaxies. Grand majestic spirals like the Milky Way and Andromeda, huge elliptical collections of old stars, small irregular dwarf galaxies—there's an overwhelming variety, and an overwhelmingly large number, too, spread out over a volume that is many billions of light-years across.

In December 1995, for the first time, astronomers trained the Hubble Space Telescope on a tiny, seemingly empty patch of sky. They left the camera shutters open for ten days straight. The result was a breathtaking photo of more than a thousand faint, distant galaxies in an area that would disappear behind the head of a pin held at arm's length. One pinhead's diameter to the left or to the right would reveal another thousand distant galaxies.

So here's the current view of the observable universe: it's vast, dark, cold, and empty. But sprinkled throughout the void are about 2 trillion galaxies, grouped in clumps and clusters. Are you far out in space and want to find your way home? Better buy yourself an incredibly accurate navigation system—there are no road signs along the cosmic highways. Finding the proverbial needle in the haystack is a lot easier.

If you succeed in locating the Milky Way galaxy, pause for a second to take in the view. A few hundred billion suns are arranged in beautiful spiral arms amid stellar clusters, bright nebulae, and dark dust clouds. And just one of them—a rather inconspicuous middle-of-the-road star—is our Sun. It lives out its life in one of the Milky Way's quiet suburbs, at the inner edge of a spiral arm where nothing much happens most of the time.

Orbiting that tiny beacon are eight minuscule planets. One of the smaller four is Earth. On this mote of dust, over just a few recent centuries, human beings are beginning to unravel the mysteries of the universe.

Well, at least we're trying.

It's a humbling thought. *Homo sapiens* is almost impossible to find in the vastness of space. We're also newcomers on the cosmic stage.

Here's a revealing metaphor. Suppose all of the history of the universe was recorded in a fourteen-volume encyclopedia. Fourteen fat tomes, a thousand pages each, in fine print. The big bang would be on the first line on the first page of Volume 1. The first stars and galaxies would form somewhere halfway through Volume 1. But the birth of the Sun and the planets is described only in Volume 10. The dinosaurs go extinct on page 935 in Volume 14. *Homo sapiens* appears on the bottom fifth of page 1,000. All of our written history would be crammed in the second half of the very last line.

The astronomical view is one way to understand our world. Many physicists would use a different approach: rather than simply describe everything you see (galaxies, stars, planets), find out what everything is made of, and how it all works.

Suppose an astronomer and a physicist both study J. R. R. Tolkien's *The Lord of the Rings*. The astronomer presenting her findings would describe the trilogy's story line, the protagonists, the metaphorical meaning, the writing style, and so on. The physicist would describe the alphabet, letter frequencies, rules of punctuation, and grammar.

But aren't these the same for many widely differing books? "Yes!" the physicist would exclaim enthusiastically; that's the great thing about this approach. You can stop focusing on the peculiarities and start finding the common fundamentals in order to develop the deepest possible understanding. Of course, both approaches have their pros and cons. They're really very complementary.

So, just as every possible book is composed of a small number of different letters and has to abide by the rules of grammar, all objects in the universe are composed of just a small number of elementary particles that interact through the fundamental forces of nature.

The surprising thing is that the world around you—pinheads, people, planets, and protoclusters—is composed of just three types of elementary particle: the up quark, the down quark, and the electron. And just as letters can be grouped into words, sentences, paragraphs, and books, those three particles make up atoms, molecules, compounds, and literally every object you can think of.

As for the fundamental forces of nature, physicists know of only four. Two of them act at very short range—they play a role only at the scale of an atomic nucleus. That's why they're called the strong and the weak nuclear force. The other two—electromagnetism and gravity—can be experienced in the world at large, as everyone knows who has switched on a lightbulb or dropped a wineglass.

I'm leaving out lots and lots of details here. Neutrinos, unstable elementary particles, antimatter, force-carrying particles, the famous Higgs boson, dark matter, supersymmetric particles, tetraquarks, a possible fifth force—the list goes on and on. If you're interested, you can pick up a popular book on particle physics, so I won't go into the subject at any length, though I'll get back to neutrinos and dark matter later on in this book.

One particular detail is important to our story about spacetime and gravitational waves: the weirdness of gravity. We're all quite familiar with its obvious effects. But somehow, gravity behaves very differently from the other fundamental forces of nature. According to Albert Einstein, that's because gravity is intimately connected to space and time.

Now try to explain that to Isaac Newton. Newton never knew the true nature of gravity, of course. He just derived a universal formula that usefully described the attractive force between two masses at a certain distance from each other. But like most of his contemporaries, Newton considered space and time to be independent, absolute concepts.

In fact, Newton's views on space and time are much like our own intuitive ideas. Space is simply there—a three-dimensional nothingness that goes on forever. An object (like an elementary particle or a planet) can sit at a certain location in space, or it can move from one location to the next. If we choose a particular reference point, all other locations can be identified by just three coordinates. Starting at your reference point, the three numbers tell you how far you have to go forward or backward, left or right, and up or down to reach the other location. Space is a bit like three-dimensional graph paper. It's the empty, unchangeable backdrop against which all events in the universe play themselves out.

And time? The imaginary clock of nature ticks away the moments that make up a dull day as well as all the seconds since the birth of the universe. Time is the absolute, infallible metronome of the cosmos, tagging each and every event with a unique time stamp. Oh, and it's one-dimensional: if you choose a reference moment, you need only one number to tell you at what time any other event takes place.

I'm confident you have no trouble imagining space and time the way Newton did. It's the natural way to think of them. Our brains are wired to come up with this convenient view.

Alas, it's wrong.

What Einstein showed is that space and time are linked. Three-dimensional space and one-dimensional time are actually interwoven in four-dimensional spacetime.

Einstein also showed that space and time are not absolute, but relative. That's of course why his revolutionary theory is called the theory of relativity. What's the distance between two points in space? Depends on whom you ask. For someone traveling at half the speed of light, the distance between two points in space is much smaller than for someone at rest (at rest with respect to the two points, that is). The same is true with the time interval between two events. The faster you go, the slower your clock is ticking. The only thing that is absolute—the same for all observers, independent of their motion—is the four-dimensional separation between two events (at two locations) in spacetime.

Finally, Einstein showed that mass (and energy, too) exerts an influence on four-dimensional spacetime. Straight lines are slightly bent under the influence of massive objects such as stars or black holes. (For smaller and lighter objects—asteroids or apples—the effect is completely negligible.) The result: anything that follows a straight line, like a ray of light or a planet, starts to move along a curved path in the presence of a massive body. What we perceive as gravity is really the effect of the curvature of spacetime on the motions of other bodies. And since we're talking about the curvature of space*time,* time is also influenced by the presence of massive bodies—close to a black hole, your clock starts to tick slower and slower.

If you think this all sounds crazy, go ask the fictional *Interstellar* astronaut Joe Cooper. Together with his crewmates Amelia Brand and Doyle, he spends just a few hours on a world known as Miller's planet, which orbits the giant black hole Gargantua. Because the planet orbits so close to the black hole, spacetime is curved dramatically, and time there proceeds at a snail's pace. When Cooper, Brand, and Doyle return to *Endurance,* Nikolai Romilly, the fourth crew member, has aged twenty-three years.

The strong spacetime curvature is also evident in the appearance of Gargantua itself. The black hole is surrounded by a flat, equatorial disk of superheated gas from which matter falls into the hole. Normally, you would expect to see only the near side of the disk. After all, the far side is behind the black hole. But thanks to the curvature of spacetime, light from the far side is bent all the way around Gargantua. As a result, the black hole appears to be surrounded by a bright ring.

At times, I imagine, Kip Thorne's involvement must have been annoying to the visual effects artists and computer animators at Double Negative, the firm in London that had to turn his spacetime equations into breathtaking footage. Sometimes the Caltech physicist wasn't allowed the last word, and scientific accuracy had to be compromised; as Thorne describes in his 2014 book *The Science of Interstellar*, movie director Christopher Nolan didn't want to confuse his audience *too* much. But in the end, Thorne was extremely satisfied. "What a joy it was when I first saw these images!" he writes. "For the first time ever, in a Hollywood movie, a black hole and its disk depicted as we humans will really see them when we've mastered interstellar travel."

So we can describe and visualize the effects of spacetime curvature on the bending of light paths and the flow of time. But how should we imagine this four-dimensional construct, let alone its curvature?

In 1917, Albert Einstein wrote a short book about his new theory simply called *Relativity: The Special and the General Theory*. Later, others wrote about relativity, too. One of the funniest is *Mr. Tompkins in Wonderland* (1940), written by cosmologist George Gamow. It's still in print, for good reason. As a young teenager I devoured yet another book called *A Guided Tour through Space and Time* (1959) by Hungarian physicist Eva Fenyo. And if you really want to dive into the topic, you should read Kip Thorne's impressive 1994 book *Black Holes and Time Warps: Einstein's Outrageous Legacy*. It's over six hundred pages, but it's written for the general reader.

The trick everyone uses to visualize four dimensions is quite straightforward: forget about one of them. Of course, we don't want to ignore the time dimension. But it's okay to throw one space dimension overboard. That leaves us with two dimensions of space and one of time. As a result, spacetime has become three-dimensional—something we're familiar with.

In two-dimensional space, things can move only forward or backward and left or right. There's no up or down. So let's focus on motions that occur in two dimensions, on a horizontal plane.

I want you to visualize two things that travel through this plane in a straight line. One is a ray of starlight, traveling at 300,000 kilometers per second. The other is a planet, traveling in the same direction, but ten thousand times slower, at a mere 30 kilometers per second. If there's no external influence on either of them, both will follow the same straight path, albeit at very different velocities.

Let's now put the Sun in the same plane, at some 150 million kilometers from the straight path. We know that the mass of the Sun creates a curvature in spacetime. As a result, both the path of the light ray and the path of the planet are bent. But something strange is going on: the path of the light ray is bent by only a very small amount— we'll come back to the Sun's light-bending effect in Chapter 3. But the path of the planet (let's call it Earth) is bent much more strongly—all the way into a circular orbit. What's happening? If both are influenced by the same curvature, wouldn't you expect both of them to follow the same curved path?

No, you shouldn't. Here's the reason: we're talking about curved paths not in space but in space*time*. If we really want to understand what's going on, we should add the time dimension to our two-dimensional space and consider motion in three-dimensional spacetime. Here, time has taken the place of the third space dimension (up / down). In fact, we've created a new three-dimensional coordinate system. The *x* axis and the *y* axis—in the horizontal plane—have tick marks every 300,000 kilometers (the distance light travels in one second). The vertical *z* axis has similarly spaced tick marks every second.

Let's look again at the light ray. At time zero, it is located at a certain point in space. One second later, it has traversed 300,000 space kilometers—one tick mark in the horizontal plane. But in three-dimensional spacetime, it has also traveled one tick mark up. After all, one second has passed. So in spacetime, the light ray moves along a path that is tilted at 45 degrees.

Now, let's look at Earth. In one second, it travels just 30 kilometers. It takes our planet 10,000 seconds to cover 300,000 space kilometers (that's two hours and almost forty-seven minutes). So the path of Earth in three-dimensional spacetime (its *world line*) is much less tilted than the path of the light ray is: just 20 arc-seconds or so (an

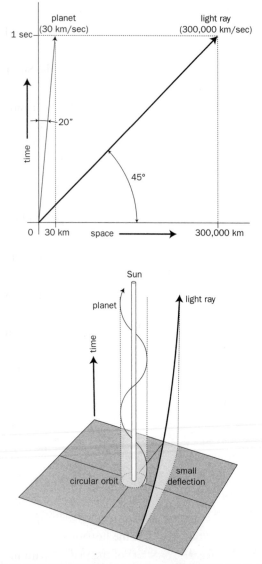

In spacetime, the "world line" of a light ray, moving at 300,000 kilometers per second, has a tilt of 45 degrees, while the world line of a planet moving at just 30 kilometers per second is almost vertical *(top, angles not to scale)*. Both world lines are only slightly bent by the spacetime curvature produced by the Sun *(bottom)*, but if projected into space coordinates *(horizontal plane)*, the planet appears to be deflected much more strongly than the light ray is.

arc-second is 1 / 3,600 of a degree). To a casual observer, the light ray obviously moves diagonally while the planet moves nearly straight up—*almost* vertically.

So far so good. Now what happens if we add the Sun to the picture? In our simplified story, the Sun doesn't move in space—its velocity is 0 kilometers per second. Consequently, it moves exactly vertically in three-dimensional spacetime. But the Sun's mass creates a tiny curvature in spacetime. As a result, the world line of the light ray and the world line of the planet are both bent very, very slightly. Here's how.

The diagonal world line of the light ray is slightly bent, but not for long because of its high velocity. In no time, the light ray has left the area where spacetime is curved by the mass of the Sun. As before, it travels along a straight spacetime path, at the same vertical tilt of 45 degrees, although the tilt is now in a slightly different direction. Projected back onto the plane of two-dimensional space, we see that the light ray's path has experienced a slight change of course.

Earth, however, stays in the curved region. It keeps moving through spacetime along an almost vertical path at the same angle of 20 arc-seconds. But the direction of this small tilt is slowly changing all the time because of the curvature induced by the mass of the Sun. After almost 8 million seconds (three months or so), the direction of tilt has changed by a full 90 degrees. Projected onto the plane of two-dimensional space, we see that the planet has completed one-fourth of an orbit around the Sun.

But that's not a strong curvature at all! In 8 million seconds, the planet has moved 8 million tick marks "up" through spacetime. In the same time, it has traversed a mere 236 million space kilometers. That's less than 800 tick marks in the horizontal plane. It would be extremely hard to see the curvature of the planet's path in spacetime by eye—it's still almost a perfect vertical line.

After one year, Earth has completed a full orbit around the Sun, corresponding to just over 940 million space kilometers. But it has taken 31.5 million seconds to do that. Earth's helical world line in spacetime is almost indistinguishable from a straight line. That's because the Sun is not extraordinarily massive, and the resulting curvature of spacetime is small. Still, if we forget about the time dimension and look only in the plane of two-dimensional space, Earth's path is strongly curved, all the way into the familiar circular orbit. Our speeding light ray, meanwhile, has traveled almost a quarter of the distance to the nearest star.

―――――――

This is all quite difficult to grasp if you are hearing about it for the first time—and I didn't even ask you to visualize *four*-dimensional spacetime. (If you've lost track, you can always reread the previous pages tomorrow morning, or next week.) In any case, you now understand why our everyday intuition lets us down if we want to understand the peculiarities of spacetime and general relativity.

And that's a good lesson. In dealing with colliding black holes, extremely strong spacetime curvature, and gravitational waves, we can't trust our gut feeling. Instead, we have to rely on supercomputer calculations based on Albert Einstein's theory of general relativity. If we trust Einstein, we'll have to accept the results of those calculations.

This is one of the reasons why Kip Thorne was so happy with the *Interstellar* project. A visual effects company such as Double Negative has much more powerful computers at its disposal than a Caltech theoretical physicist does. The resulting movie sequences provide scientists like Thorne with new, valuable insights. As he writes in *The Science of Interstellar,* "For me, those film clips are like experimental data: they reveal things I never could have figured out on my own, without those simulations."

Now what do scientists do when they gain new insights? Publish a paper, of course. And that's just what Thorne did. Two papers, actually—one on *Interstellar*'s wormhole, the other on the movie's giant black hole, Gargantua. You can look them up on the Internet. The first one, titled "Visualizing *Interstellar*'s Wormhole," was published in the prestigious *American Journal of Physics.* The other paper, "Gravitational Lensing by Spinning Black Holes in Astrophysics, and in the Movie *Interstellar,*" appeared in another professional journal, *Clas-*

―――――――

sical and Quantum Gravity. Both papers are coauthored by Oliver James, Eugénie von Tunzelmann, and Paul Franklin. James is chief scientist at Double Negative, von Tunzelmann is the company's computer graphics supervisor, and Franklin is Double Negative's cofounder and visual effects supervisor. It's nice for a theoretical physicist to be listed as an executive producer in the Internet Movie Database (IMDb), but it's equally nice for special effects experts to end up at arXiv.org—the largest electronic repository of physical science papers in the world.

———

There was just one small disappointment that Thorne had to swallow. Originally, he had hoped that gravitational waves would be featured in *Interstellar*—after all, he was cofounder of the LIGO project, and for all he knew, the first direct detection of these elusive spacetime ripples might occur in the same year the movie was released. Unfortunately, Christopher Nolan believed they would make the story line way too complicated. In any case, the first detection of gravitational waves—GW150914—didn't occur until 323 days after the official release date of the movie.

For all I know, Kip Thorne may be working on a sequel.

(((2)))

Relatively Speaking

Leiden is a poetic city.

On the side of the house at Nieuwe Rijn 36 is a poem by E. E. Cummings painted in large letters; the whole poem is some seven meters tall. It begins:

> *the hours rise up putting off stars and it is*
> *dawn*
> *into the street of the sky light walks scattering poems*

Not sure what it means, but it sounds great.

Cummings's poem is not alone: it's number twenty-three in a series. There are about a hundred other wall poems in the historic center of Leiden, which is a bit more than forty kilometers south of Amsterdam, the capital of the Netherlands.

One poem stands out from the rest. It's painted on the eastern wall of Museum Boerhaave, the Dutch National Museum for the History of Science and Medicine. But it's hard to read out loud. It's in a language not many people are familiar with, and it's just one line:

$$R_{\mu\nu} - \frac{1}{2}R\,g_{\mu\nu} + \Lambda\,g_{\mu\nu} = \frac{8\pi G}{c^4}T_{\mu\nu}$$

Perhaps it doesn't seem like a poem. This is the field equation of Albert Einstein's theory of general relativity. You'll note that the equation consists of two parts, separated by an equals sign, which means that the part on the left equals the part on the right. The left part describes the curvature of spacetime. The right part describes the distribution of mass (and energy). Change the distribution of mass, and you'll change the curvature of spacetime. Change the curvature, and matter will start to move around—as we saw in Chapter 1.

Einstein's field equation is written in the language of mathematics. But the best translation into English is by John Archibald Wheeler, a brilliant American physicist (who was also Kip Thorne's mentor, by the way): "Matter tells spacetime how to curve; spacetime tells matter how to move." Poetry after all.

The equation on the wall of Museum Boerhaave was painted to commemorate the centenary of Einstein's theory, and it was unveiled in a ceremony in November 2015 by Dutch physicist Robbert Dijkgraaf, the director of the Institute for Advanced Study in Princeton, New Jersey, where Einstein worked the last twenty-one years of his life—quite fitting.

From Museum Boerhaave, it's a fifteen-minute walk to the Boerhaave's storehouse at Raamsteeg 2. Paul Steenhorst, the museum's head of restoration, has something to show me. He takes me one flight up to room N1.01, a fully climate-controlled room that houses the physics collection in a series of pine cabinets. Paul opens drawer J410 and takes out item V34180. It's a small navy blue cardboard box. "Waterman's Ideal Fountain Pen" is written on the lid.

A minute later, I'm holding Albert Einstein's fountain pen, the one he used for *everything* he wrote between 1912 and 1921, including the original manuscripts of his 1915 papers on general relativity. The curvature of spacetime, the field equations, gravitational waves—it all flew from this delicate *Füllfeder*, as Einstein would've called it.

Does the phrase "six degrees of separation" ring a bell? It's the notion that says you're at most six links away from any other person on Earth. A fountain pen is not a person, but in a sense I'm only two links away from the greatest physicist who ever lived.

By the way, I'm not making up this characterization. Einstein really is considered to be the greatest physicist ever. At least, that was the result of a 1999 poll among a hundred prominent scientists carried out by the magazine *Physics World*. That same year, *Time* magazine named Einstein the Person of the Century—not just physicist, mind you, but person.

———

Everyone knows who Albert Einstein is. Big mustache, uncombed hair, sloppy sweater, sandals—he has become the iconic scientist. Not many other physicists have their face immortalized on postcards, coffee mugs, and T-shirts. Yes, it helped that he poked out his tongue for UPI photographer Arthur Sasse on his seventy-second birthday. But it was really his genius that skyrocketed him to scientific stardom.

You may be surprised to learn that you know a lot more about the universe than Einstein did when he penned his theory of general relativity. Back then, no one had seen the far side of the Moon. Pluto had yet to be discovered. Astronomers didn't know the energy source of the Sun. The true nature of spiral nebulae—galaxies such as our own Milky Way—was unclear. Most scientists thought the universe had existed forever. The discovery of pulsars, quasars, and exoplanets was still decades away. Antimatter, neutrinos, quarks—in 1915, those words would have meant nothing to Einstein. The same would be true for galaxy clusters, gamma-ray bursts, and dark matter.

What scientists *did* know in 1915 was that the universe is ruled by gravity, despite the fact that gravity is a tremendously weak force. The electromagnetic force, for instance, is much stronger. But electromagnetic forces can be either positive or negative—attractive or repulsive. In the universe at large, these opposite forces cancel each other out. Gravity, however, is always attractive (antigravity hasn't yet escaped the realm of science fiction). As a result, the motions of stars and planets—and of tripping people and falling apples, of course—are governed by this one feeble force.

In case you doubt what I said about gravity being a very weak force, here's a simple experiment to prove me right. Tear a small piece of paper

Albert Einstein in Princeton, New Jersey, in 1947.

into shreds and let them drop onto your desk. They flutter down because of the gravity of Earth—the same force that keeps you from gently floating toward the ceiling. Now take a small plastic comb and rub it through your hair or on your woolen sweater. Then hold the comb a few centimeters above the snippets of paper on your desk. See what happens? They are immediately attracted by the static charge of the comb. There you go: the electromagnetic force of a statically charged pocket comb is much stronger than the gravitational force of a whole planet! Which means gravity is a truly weak force of nature.

The Greeks didn't know much about electromagnetic forces (and nothing about the strong and weak nuclear forces, for that matter). They also didn't know much about gravity. Aristotle believed that all objects had a natural tendency to move toward the center of the universe. He also believed that the center of the universe was occupied by the Earth. Therefore, things fall down—it was as simple as that. Moreover, Aristotle was convinced that heavy things fall faster than lightweight things. Who knows—maybe he experimented with snippets of vellum and Greek amphoras.

Too bad Aristotle never saw the footage of Apollo 15 commander David Scott, who dropped a feather and a hammer on the Moon (you can easily find the video on YouTube). The Moon has no atmosphere, so there's no air resistance. And without air resistance, the feather falls at exactly the same pace as the hammer—it looks weird. (And both fall six times slower than the hammer would on Earth because the Moon has only one-sixth the gravity that we're used to.)

Legend has it that Galileo Galilei pioneered a similar experiment in 1589 from the Leaning Tower of Pisa in Italy. It's pretty straightforward. Take two spheres of different weight—maybe one made of lead and the other of wood. They should be large and heavy enough to avoid being affected much by air resistance. Climb the tower. Let the two spheres drop at exactly the same time. Which one lands first? If they hit the ground simultaneously, you've proven Aristotle wrong.

However, there's no reliable account of Galileo actually performing the experiment. Yes, he describes it, but probably as a thought experiment. And if he really *did* drop spheres from a tower, he was certainly not the first to do so. In 1585, Flemish scientist and mathematician Simon Stevin and his friend Jan Cornets de Groot (who would later become the mayor of the Dutch city of Delft) carried out the experiment from the tower of the New Church in Delft. It's documented in detail in a book Stevin published in 1586. I really like the Stevin story—the New Church is very near where my father was born.

Anyway, by the end of the sixteenth century, Aristotle was proven wrong—finally. (As you read in Chapter 1, Aristotle's idea about the central position of the Earth had already been rejected a couple of decades earlier by Copernicus.) But neither Stevin nor Galileo had a much better notion about the nature of gravity than the Greeks had. For example, just like Aristotle, they never considered the idea that the motions of stars and planets in the universe would be governed by the same force as the motions of leaden spheres and apples here on Earth. It took another couple of decades before Isaac Newton realized that. (By the way, the story that Newton was hit on the head by an apple that fell from a tree is also a legend.)

Newton published his ideas on gravitation in the summer of 1687, not in a scientific paper but in a comprehensive three-volume book in Latin called *Philosophiae Naturalis Principia Mathematica* (*Mathematical Principles of Natural Philosophy*). The first English edition didn't appear until 1728, more than a year after the author's death. Less than two centuries after the publication of the *Principia,* on March 14, 1879, in Ulm (present-day Germany), Pauline Einstein-Koch gave birth to her first child, Albert. Albert Einstein would prove Newton wrong.

You've heard a legend about Galileo Galilei. I mentioned one about Isaac Newton. There are enough legends about Albert Einstein to fill this book. Fortunately, the true story of his life is at least as exciting as the myths. Just as legendary, you might say.

Albert was only one year old when his Jewish parents moved from Ulm to Munich. His father, Hermann, ran a small factory producing electrical equipment together with Hermann's brother. His mother took care of the house and the family, and in November 1881, Albert's baby sister, Maja, was born. Every now and then, his aunt Fanny (his mother's sister) would visit with her daughters Hermine, Elsa, and Paula. Young Albert grew up among women; he was very fond of his sister, and he loved to play with his cousin Elsa.

Was he a special child? Not particularly. Yes, he was quiet and introverted. At a young age, he learned to play the violin. He was good at it, too. And he was fascinated by things other children wouldn't care too much about, like the compass his father gave him at age five. Whichever way you rotate the compass case, the needle points in the same direction. Apparently it is influenced by something in space itself—amazing! But Hermann could never have imagined his son would become the greatest scientist of all time.

Albert's father had other worries. His company went bankrupt in 1894. The family moved to Milan—maybe luck would be on his side in Italy. At the time, fifteen-year-old Albert was attending the Luitpold Gymnasium in Munich, so he stayed behind. By then, he had developed a strong interest in physics. His goal was to study at the renowned Swiss Federal Polytechnic School in Zurich.

Albert also had a strong interest in girls. (As I said, he wasn't *that* special—most teenage boys have a strong interest in girls.) Girls also had a strong interest in Albert. He was a handsome guy with curly black hair and beautiful dark eyes. Marie Winteler was one of the girls who were enchanted with him. Marie was the daughter of ornithologist Jost Winteler, a teacher at the Argovian Cantonal School in Aarau, Switzerland. Albert stayed at the Winteler residence during his two-year period of study in Aarau. Pretty soon he and Marie fell in love.

In September 1896, Albert passed the school's exam with impressive grades, at least in the physical sciences. "Don't know much about history . . . don't know much about the French I took"—some of the lyrics to Sam Cooke's 1960 hit single "Wonderful World" could well have been written by Einstein. But when it came to physics, algebra, and geometry, he obtained the highest possible score. At age seventeen he was admitted to the Polytechnic.

Could a seventeen-year-old teenager imagine he'd be the one who would solve a number of nagging problems in physics? I doubt it. But Albert Einstein was certainly *aware* of those problems. One particular enigma had been around for decades, and it could potentially spell bad news for Newton's theory of gravity.

The nice thing about Newton's theory is that it allowed astronomers to finally understand the motions of the planets in the solar system. Using Newton's equations, it was relatively easy to predict where a planet would be, say, twenty years from now. Or to describe where a planet had been half a century ago—it's basically the same kind of calculation.

I say "relatively easy" because the solar system is a complicated scene to begin with. If it was just one Sun and one planet, applying Newton's equations would be a piece of cake. In reality, a planet's motion is also slightly influenced by the gravity of the other planets in the system. To predict the path of Saturn, for instance, you have to take the gravitational force of Jupiter into account. Sometimes Saturn will be slightly slowed down by Jupiter's gravity; at other times it will be slightly speeded up. Calculating all those disturbances is not a piece of cake; it's a complete bakery.

A vital test for Newton's theory presented itself in 1781. That year, English astronomer William Herschel discovered a new planet beyond the orbit of Saturn: Uranus. Immediately astronomers used Newton's equations to predict the new planet's future path. Of course they took the gravity of the other major planets into account. But quite soon it became apparent that Uranus was slowly veering off its predicted course. Could Newton's theory of universal gravitation be wrong after all? Or might there be yet *another* planet, pulling Uranus this way and that?

In the 1840s, mathematicians turned Newton's equations around. Normally you would know the locations of all the planets, and that enables you to precisely calculate their orbits. But what if you could do the math backward? In that case, you'd start with the deviant orbit of Uranus and calculate where you might find the unknown planet that's responsible for the deviations. French mathematician Urbain Le Verrier took up the challenge.

Today it would be easy to develop computer software to solve the puzzle—it's something every astronomy student should be able to do in a day or so. But this was the time of writing desks, pencil and paper, and logarithm tables. It took Le Verrier months to arrive at a reliable answer.

His efforts paid off. In September 1846, a new planet was found close to the position calculated by Le Verrier. He had written a letter with his predictions to his colleague Johann Galle at the Berlin Observatory. Within hours, Galle and his assistant Heinrich d'Arrest had found Neptune, as it was called.

You'll now understand why Neptune is sometimes called the "writing desk planet": it was found on the basis of mathematical calculations. Those calculations made use of Newton's equations. So the discovery of Neptune, the eighth planet in the solar system, was seen as a triumph for Newton's theory of universal gravitation.

This is indeed how science normally works. It starts with observation—in this case, the motions of falling apples and planets. Some genius comes up with a theory that neatly explains the observations—in this case, Isaac Newton and his theory of universal gravitation. As more and more predictions of the theory are confirmed, scientists become ever more confident of the theory's validity—that's how Neptune validated Newton.

Some ten years after the discovery of Neptune, Le Verrier began searching for a ninth planet. Not beyond the orbit of Uranus, but within the orbit of Mercury, the innermost planet in the solar system. The reason: just like Uranus, Mercury misbehaved.

Mercury's path around the Sun is not a perfect circle. It's definitely eccentric: the distance to the Sun varies during each orbit. Moreover, the orbit itself slowly rotates—Mercury's closest point to the Sun (its perihelion) shifts over the course of time. In the mid-nineteenth century, this so-called perihelion precession had been measured quite precisely: about one-sixth of a degree per century, which turned out to be larger than Newton's theories predicted. As Le Verrier calculated, 92.5 percent of Mercury's perihelion precession could be attributed to the gravitational disturbances of the other planets. But 7.5 percent (43 arc-seconds per century) remained unaccounted for. The discovery of Neptune didn't help—Neptune is too far away and moves too slowly to have a noticeable effect on Mercury's orbit.

So Le Verrier proposed that there must be another as-yet-undetected planet within the orbit of Mercury. Could such a nearby planet have escaped detection? It certainly could. A planet that's really close to the Sun would rise and set almost simultaneously with the Sun. As a result, it's only up in the sky during the day, when you can't see it. There are just two rare occasions when it might be visible. The first is during a total solar eclipse, when the bright light of the Sun's disk is obscured by the Moon. The second is during a transit, when the planet happens to pass in front of the Sun's disk, as seen from the Earth.

Because Le Verrier had successfully predicted the existence of Neptune based on the misbehavior of Uranus, he was convinced that the precession of Mercury's orbit could also be explained by a previously unknown "intramercurial" planet. Le Verrier even coined a name for his hypothetical Sun-hugging planet: Vulcan, after the Roman god of fire.

The problem was that no one ever found Vulcan—not during eclipses, not during its expected transits. (We now know for sure it doesn't exist.) So at the close of the nineteenth century, when Albert Einstein started to study physics and mathematics in Zurich, he realized Newton's theory of universal gravitation was in trouble: it could not fully explain the slow precession of Mercury's orbit. What could possibly be wrong?

Young Albert was aware of another pesky problem, too. It had to do with the speed of light.

Light moves incredibly fast. So fast, in fact, that scientists had trouble measuring it. To give you an idea, if someone switches on a laser pointer in New York, its light would take only 0.013 seconds to reach Los Angeles (if the curvature of the Earth weren't in the way). It wasn't until the second half of the seventeenth century that Danish astronomer Ole Rømer arrived at a good estimate of the speed of light. We now know it's about 300,000 kilometers per second. (Actually, it's 299,792.458 kilometers per second in the vacuum of empty space. We've just been awfully lucky in choosing our metric units in such a way that the speed of light turns out to be so close to a nice round number. In other units, the value for the speed of light would be harder to remember. For instance, it's 670,616,629 miles per hour or—for my older British readers—1.803 trillion furlongs per fortnight.)

Just fifteen years after Rømer's experiments, in 1690, Dutch physicist Christiaan Huygens published his famous book *Treatise on Light.* Huygens was one of the greatest scientists of his time. He discovered the true nature of the rings of Saturn. He found Saturn's largest moon, Titan. He was the first to spot dark marks on the surface of Mars. He much improved the study of mechanics and optics, and he invented the pendulum clock.

In *Treatise on Light* (originally published in French), Huygens argued that light is a wave phenomenon. Think of it as a propagating wave on the surface of a pond. Just like water waves or sound waves (and, as we will see, gravitational waves), light waves are characterized by a number of properties. So it's a good idea to start looking at the general properties of any kind of wave.

First, there is the wave's *amplitude.* For water waves, the amplitude is just half the difference in height between crests and troughs. For sound waves or light waves, the amplitude is a measure of energy—the sound's volume or the light's brightness. For gravitational waves, the amplitude is the strength of the wave: more powerful waves have a greater effect on the curvature of spacetime.

Next, there is the *velocity* of the wave. A ripple on a pond propagates at something like 1 meter per second. Sound waves in air propagate at about 330 meters per second. Light waves—and gravitational waves, too—move at the velocity of light, some 300,000 kilometers per second.

Finally, there is the wave's *frequency.* This is simply the number of wave crests you see passing by each second from a stationary point. Float a rubber duck on a pond, and a water wave's frequency tells you how fast the duck bobs up and down. If the wave crests are close together—that is, if the wavelength is small—the frequency will be relatively high, and the duck will bob quickly. Longer wavelengths, where the waves are very stretched out, correspond to lower frequencies, and a slower bobbing motion.

From looking at the world around us, it seems obvious that a propagating wave requires a medium to propagate through: ripples on a pond propagate through water, and sound waves propagate through air. Little wonder, then, that scientists came up with the idea of the ether—a mysterious substance that would fill all of empty space. The ether would be the medium through which light waves propagate.

But the problem for physicists at the close of the nineteenth century was that they could find no evidence for the existence of the ether. If there was such a substance, our planet would be moving through it in various directions while orbiting the Sun, and so Earth would have its own velocity with respect to the ether. And that velocity should be evident in measurements of the speed of light.

Here's why. Imagine light from a distant star propagating through the ether at 300,000 kilometers per second. Earth's orbital velocity around the Sun is almost 30 kilometers per second. So if Earth moves "upstream," in the direction of the star, you would expect the light waves to arrive at a relative velocity of 300,030 kilometers per second. If we're moving "downstream," in the same direction as the light waves, you would expect to clock them at 299,970 kilometers per second. (It becomes a bit more complicated if the solar system is also moving through the ether, but you get the picture.)

Enter American physicists Albert Michelson and Edward Morley. In the spring of 1887—Albert Einstein had just celebrated his eighth birthday—they set up a delicate experiment in Cleveland, Ohio. There's no need to go into the details of the experiment, but it's fun to know that they used an interferometer, the same type of instrument that made the first direct detection of gravitational waves in September 2015.

Michelson and Morley's device was sensitive enough to measure small differences in the speed of light in various directions. But they found none. In whichever direction they looked, light waves always moved at the same velocity: 300,000 kilometers per second. It was as if Earth dragged the hypothetical ether with it as it moved through space. At the time, no one had a satisfying explanation for this observation.

So Einstein knew that there were two observations that could not be explained by any of the theories then current: the excess perihelion precession of Mercury's orbit and the constancy of the speed of light.

There was one solution: his theory of relativity.

———————

In the fall of 1896, Albert Einstein, age seventeen, enrolled in a four-year math and physics program at the Zurich Polytechnic. At first he kept in touch with his girlfriend, Marie. Then he met Mileva Marić, and everything changed. Mileva, a Serb, was the only female student in Albert's cohort. Like Marie, she was a couple of years older than Albert. Unlike Marie, she understood the many intricacies of physics. She and Albert fell in love.

Four years later, Albert completed the program and was awarded a diploma, qualifying him to teach math and physics at secondary schools. But, rather than teach, he wanted to start work on his PhD thesis, preferably in Leiden, the Netherlands. Leiden University was home to Hendrik Lorentz, one of the greatest physicists of his time and much admired by Einstein. Lorentz's work would form the foundation for Einstein's ideas on relativity.

In 1901, hoping to get close to Lorentz, Einstein applied for a job at the Leiden low-temperature-physics laboratory of Heike Kamerlingh Onnes, another giant of science. But Kamerlingh Onnes didn't even bother to write back—a loss not just for Einstein but also for Dutch physics. Instead, Einstein eventually ended up as a patent clerk at the Federal Office for Intellectual Property in Bern, Switzerland. The father of his friend and classmate Michele Besso kindly arranged that position for him. It wasn't a very exciting job, but the quiet days at the patent office provided him with more than enough spare time to work on his physics theories.

Meanwhile, fate hadn't been too kind to Albert. Mileva had unintentionally become pregnant by Albert in the spring of 1901. Their daughter, Lieserl, was born the following January, but further details about the little girl are still unknown. Einstein biographers didn't even know about Lieserl's existence until 1986. She may have been mentally disabled, and she probably died of scarlet fever in the fall of 1903, a year after the death of Albert's father, Hermann (though some people believe that Lieserl was adopted by a friend of Mileva's and that she was alive well into the 1990s). In any case, it seems that Einstein never saw his daughter.

Albert and Mileva married in Bern in January 1903; their first son, Hans Albert, was born in May 1904. Einstein didn't involve himself too much in raising his child, let alone running the household: at the time, those were considered women's tasks, and so Mileva gave up her physics aspirations. Meanwhile, Albert embarked on his quest to solve the mysteries of Mercury's orbit and the invariable speed of light.

It was a two-stage process. The year 1905 saw the birth of the theory of special relativity. Building on the work of his former professor Hermann Minkowski, who developed the concept of four-dimensional spacetime, Einstein showed that both space and time are relative concepts. What is the distance between two points? Depends on whom you ask. The same is true for the timing of events. Two observers moving with respect to each other will arrive at different answers. And they would both be right. Bye-bye, Newton—there's no such thing as absolute space or absolute time.

Special relativity is not a simple theory. To fully understand its potential, you need to master complicated transformation equations, as they are called. But the upshot is easy to grasp. Travel at a substantial fraction of the speed of light, and bystanders will see your spaceship shrink—it's getting shorter along the direction it's traveling. That's called Lorentz contraction. Moreover, when you move fast enough, people back home will see your clock slow down. That's time dilation. The only reason we don't notice these effects in daily life is that light moves so fast. Even a Formula One racing driver isn't going fast enough to be noticeably affected by Lorentz contraction or time dilation.

One of the basic assumptions of special relativity is that the speed of light itself is the same for any observer, independent of her own motion or velocity. That is just what Michelson and Morley had observed, and Einstein took their result at face value. From Einstein's equations, it then follows that nothing can go any faster than light—it's nature's ultimate and fundamental speed limit.

In a second 1905 paper, Einstein derived his well-known equation $E = mc^2$, without doubt the most famous equation ever. It says that energy (E) can be converted into mass (m) and vice versa. It's an inescapable consequence of special relativity, and it's also closely tied to the velocity of light (c). By the way, our lives depend on the validity of this equation. As we'll see in Chapter 5, the Sun shines because of the conversion of mass into energy—something Einstein didn't know at the time—and none of the life on Earth, including us, would be possible without the Sun's energy.

Two additional 1905 papers dealt with other topics. One was on the motion of molecules, the other on the existence of photons, or particles of light. This last paper won Einstein the 1921 Nobel Prize in Physics. All in all, 1905 was Einstein's "miracle year"—he also received a PhD at the University of Zurich. He was just twenty-six years old.

The second stage of Einstein's quest was the development of the theory of general relativity. By *general* Einstein meant that it would work under all circumstances, not just in the special case of uniform, linear motions. General relativity deals with accelerated motion. This occurs when some kind of force (such as gravity or the punch of a rocket motor) causes a change of velocity or a change of direction. The theory took Einstein ten years to complete. In those years he moved from Bern to Zurich, from Zurich to Prague, from Prague back to Zurich, and then on to Berlin. In those years his second son was born (Eduard, in 1910), he wrote a heartbreaking love letter to his first girlfriend, Marie (while Mileva was pregnant with Eduard), and he fell for the charms of his cousin Elsa. In fact, when Einstein moved to Berlin in 1914, the year World War I broke out, Mileva and the boys stayed in Zurich, and Albert lived with Elsa and her two daughters, Ilse and Margot.

By then Einstein had become a well-respected physicist. In 1911, he had finally met Hendrik Lorentz during his first visit to Leiden.

He had been offered a position at Utrecht University, also in the Netherlands. He went to Prague instead, where he met and befriended the Austrian-born physicist Paul Ehrenfest in 1912. Around that time, he started to use the Waterman fountain pen that I briefly held in the Museum Boerhaave's storehouse. And in Berlin, he became professor of theoretical physics at Humboldt University, head of the newly established Kaiser Wilhelm Institute for theoretical physics, and (in 1916) president of the German Physical Society.

General relativity is a new theory of gravity. That may sound odd, but it isn't. It all boils down to Einstein's so-called equivalence principle, which he first stated in 1907. According to this principle, there's really no difference between gravity and accelerated motion.

Suppose you've just stepped into a windowless room. You're pulled down against the room's floor by the gravity of Earth. Now imagine your friend steps into a similar windowless room in a spaceship that is accelerated upward in empty space. There's no planet around to exert a gravitational force, but he, too, is pulled against the floor. That's because the whole room is accelerating upward, as part of the spaceship.

Einstein's equivalence principle says that there's no fundamental difference between the two situations. In other words, all possible experiments should yield the same results for both you and your astronaut friend. So if time slows down in an accelerating spaceship, it should also slow down in a strong-gravity environment. As Einstein explained to Lorentz during his 1911 visit, your watch runs a tiny little bit faster on the second floor of a building than in the basement, because on the second floor the gravitational field of the Earth is a tiny bit weaker.

Einstein wrestled with this problem a lot over the next few years. Eventually he drew upon the help of his friend and former Zurich classmate Marcel Grossmann to develop the complex mathematics needed to make progress on the matter. In the fall of 1915, he threw himself into a frenzy of intellectual activity, hardly leaving the attic of

Elsa's house at Haberlandstrasse 5—old-fashioned telephone (and fountain pen!) on his desk, a worn-out carpet on the floor, Isaac Newton's portrait on the wall. I can imagine he even temporarily refrained from flirting with his cousin.

During the month of November, Einstein completed four seminal papers on various aspects of general relativity: four-dimensional geometry; mass, energy, and spacetime curvature; the famous field equation that now adorns the wall of Museum Boerhaave in Leiden; and, finally, the correct prediction for the excess perihelion precession of Mercury's orbit. It could all be explained by the curvature of spacetime so close to the massive Sun.

Mission accomplished.

Einstein presented his papers on four successive Thursday meetings of the Prussian Academy of Sciences, on November 4, 11, 18, and 25, 1915. The third paper, with the Mercury result, was presented on the thirty-fourth birthday of his beloved sister, Maja—a double celebration. Every now and then he would pause in his reading to scribble formulas on the blackboard. Would every senior physicist in the room have been able to understand his work right away? Probably not. Did they realize that general relativity would revolutionize physics? Maybe, at least some of them. Were they impressed by the genius of their junior colleague? Almost certainly.

Albert Einstein was thirty-six.

It took another four years before Einstein became a cult figure—Chapter 3 will recount how that happened. By then, he had divorced Mileva (on February 14, 1919). He married Elsa less than sixteen weeks later. In 1920, he was awarded a guest professorship at Leiden University, and for many years he spent at least one month per year with Ehrenfest, who had succeeded Lorentz in 1912. Einstein became a foreign member of the Dutch Academy of Sciences and of the Royal Society. He won the Nobel Prize in Physics, visited New York, traveled through Asia, and became friends with Charlie Chaplin.

When Albert and Elsa returned from their third visit to the United States in early 1933, they decided they would not go back to Germany, where Adolf Hitler had become chancellor. After all, Einstein had Jewish roots. He was listed as an enemy of the Deutsches Reich. Books he had written were burned; his summer cottage in Caputh, not far from Berlin, was seized and later turned into a Hitler youth camp. After a nine-month stay in Belgium, the couple moved to England and from there back to the United States. In the fall of 1933, Einstein took up a position at the newly established Institute for Advanced Study in Princeton. A few weeks earlier, his good friend Paul Ehrenfest, plagued by depression, had committed suicide.

Albert Einstein's life ended on April 18, 1955. He died of an abdominal aneurysm in a Princeton hospital at age seventy-six. One of his last letters was to the family of his friend Michele Besso, who had died in March of the same year. "People like us, who believe in physics, know that the distinction between past, present, and future is only a stubbornly persistent illusion," he wrote. Time is relative, after all.

Einstein's handwriting can still be found in Ehrenfest's house, at Witte Rozenstraat 57 in Leiden. During visits, colleagues from all over the world were asked to sign a wall in the second-floor hallway just outside the guest room. The signatures read like a who's who of physics: Niels Bohr, Paul Dirac, Wolfgang Pauli, Erwin Schrödinger, Albert Einstein.

Not far from the Ehrenfest house, at Groenhovenstraat 18, is another wall poem, this one by Argentinian writer Jorge Luis Borges. This is how it ends:

Tu materia es el tiempo, el incesante
Tiempo. Eres cada solitario instante.

(You are made of time, the incessant
Time. You are every solitary moment.)

(((3)))

Einstein on Trial

Is $750 million a lot of money to spend on confirming something that everyone's already convinced is true? That's what NASA spent on Gravity Probe B, which in 2005 proved some of Einstein's predictions to be correct by measuring two subtle relativistic effects known as geodetic precession and frame dragging.

But when the project was initiated back in 1963, some people argued that there was so much new stuff to discover out there in space that it was a waste to spend vast amounts of money on merely confirming something that seems obvious.

Francis Everitt sighs; he's heard this argument too often. Everitt was the principal investigator associated with the Gravity Probe B experiment. In his office at Stanford University he recounts the convoluted history of the project, including the envy of some of his colleagues. In science, if you get money, you've got enemies—that's for sure.

At eighty-two, Everitt has a longer perspective on the money issue. From first conception through official debut to scientific results, the Gravity Probe B project took about half a century, which is incredibly long, even for a space science program. So if you spread the total cost over that entire period, you could think of the expenditure as just $14 million per year. That's less than 0.001 percent of NASA's 2016 budget. Moreover, quantitative tests of Einstein's ideas were few and far between. In other words, according to Everitt, Gravity Probe B was worth every penny.

Still, it sounds like there's a valid question in there: Why check on Einstein at all? He's the greatest physicist ever. Isn't everyone sure that he hit the mark with the theory of relativity?

Actually, no.

That's to say, scientists are never sure of *anything*. Tomorrow might bring new measurements that fly in the face of a pet theory, just like what happened when measurements of Mercury's orbit didn't fully accord with the predictions made by Newton's theory of universal gravitation. Remember how science works: Observations are explained by theory. Theory makes predictions. Experiments check on predictions. If they're confirmed, confidence in the theory grows; if not, something must be amiss. Tweak your theory or come up with a new one. Start doing experiments all over again. It's the scientific method.

So checking predictions is the bread and butter of science. Francis Everitt likes to quote Leonard Schiff, the Stanford physicist who came up with the original idea for Gravity Probe B: "What's the sense of a theory without experiments?"

I'll come back to Gravity Probe B, geodetic precession, and frame dragging in much more detail at the end of this chapter. First, let's go back a century or so. Albert Einstein had just formulated his theory of general relativity. It nicely explained everything we observe in the world around us: falling apples, orbiting planets, and even the excess perihelion precession of Mercury. Great. But was general relativity really the last word on gravity and spacetime? Was Einstein right?

Einstein himself came up with three ways to check his new theory. The first was whether it would successfully explain one of the observations that had spurred Einstein to begin working on his theory in the first place: the odd behavior of Mercury, the fact that its elongated orbit rotated a bit faster than Newton would've expected. And indeed, the theory fully accounted for Mercury's precession.

The other two checks, though, were based on specific predictions made by the theory of general relativity. One was the deflection of starlight. The other was gravitational redshift. Try me, said Einstein in essence. If I'm right, starlight should be deflected by massive bodies, and the wavelength of light should shift in a strong gravitational field. If nothing of the sort is found, I'm wrong and we'll have to start all over again.

Let's begin with the deflection of starlight. Imagine the Sun as seen from the Earth. The Sun has a backdrop of stars. You can't see them, of course, because the Sun is too bright, but they're there. For each and every day of the year, we know exactly what part of the sky the Sun is in.

Now imagine a ray of light from a star that is observed close to the edge of the Sun. The star's light travels through the universe along a straight line for tens or hundreds of years or more, right in the direction of our telescope. But then the light passes close to the Sun. Because the Sun is a massive body, it produces a local curvature in spacetime, as we saw in Chapter 1. As a result, the path of the light ray is bent. The light will move on in a slightly different direction and won't end up in our telescope.

But if that light doesn't end up in our telescope, do we see the star at all? Yes, of course we do. There are other rays of light from the same star that are emitted in slightly different directions into space, also traveling in straight lines. Under other circumstances those light rays would pass to one side or the other of our telescope. But upon passing close to the Sun, their paths, too, are bent by spacetime curvature, and so they end up in our scope.

This is the prediction of Einstein's theory of general relativity: we can observe starlight that has had its path bent by spacetime curvature. If there were no spacetime curvature, starlight that passes very close to the edge of the Sun would give us an image of the star just at the edge of the Sun. But because light that travels close to the Sun does get deflected onto a slightly different path than before, we see that star as a little farther away from the edge of the Sun than it really is—that is, we see the star in the "wrong" position.

In a sense, the Sun acts like a lens. It sort of magnifies the star field in its immediate vicinity. At larger apparent distances from the Sun, the effect becomes too small to observe. But close to the Sun's edge, all the stars appear to be pushed outward a tiny little bit. There you have it—the deflection of starlight caused by the curvature of spacetime.

There's an odd twist to this story. Not many people know it, but Newton's theory of universal gravitation *also* predicted starlight deflection. That sounds weird—after all, isn't light massless? How can something without mass be attracted and deflected by a massive body like the Sun? Well, imagine two things orbiting the Sun at the same distance: the Earth and an apple. The Earth is much more massive than the apple. As a result, the gravitational force experienced by the apple is much smaller than the force experienced by the Earth. But for less massive bodies, smaller forces are sufficient to generate the same acceleration. In fact, that's what Simon Stevin and Jan Cornets de Groot demonstrated when they dropped spheres of different masses from the tower of the New Church in Delft. What's true for spheres of different masses is also true for the Earth and the apple. They're accelerated to the same degree. As a result, they follow the same path around the Sun.

So in Newton's theory, gravitational acceleration is independent of mass. Apples are accelerated to the same degree as planets. Even an extremely low-mass elementary particle such as an electron experiences the same amount of gravitational acceleration. The mass of the planet, the apple, or the electron doesn't show up in the resulting formula at all. So even if the mass is really zero, as it is for light, Newton's theory predicts gravitational acceleration. (The resulting amount of deflection is of course very small, thanks to the high velocity of light.)

In 1911, Einstein made his first prediction for the deflection of starlight by the Sun. Frustratingly, he arrived at the same value as Newton had—just under one arc-second. If both theories predict the same value, then no experiment can possibly support one theory over the other. But in 1916, Einstein realized that he had made a mathematical error and that general relativity really predicted a deflection about twice as large as the Newtonian value, a whopping 1.75 arc-seconds.

In everyday life, a deflection of 1.75 arc-seconds isn't all that much. Imagine that your friend shines a flashlight in your direction from a distance of 120 meters. You carefully measure the direction from which the light is coming. Then your friend moves the flashlight by just 1 millimeter. That's a change in direction of 1.75 arc-seconds. I bet you'd have trouble measuring it.

There's another problem: the effect occurs only near the visible edge of the Sun. Ever tried to observe stars in bright daylight, let alone measure their positions? It's a bit like trying to study fireflies that hover in the distance far behind a floodlight in the foreground. You'd hope someone switches off the floodlight, or at least covers it up somehow.

Something like that was the solution to the problem of measuring starlight deflection. Every now and then, the Sun is blotted out temporarily as the Moon moves in front of it. During a total solar eclipse, the bright surface of the Sun is completely blocked, or occulted, by the Moon, and the stars in the background become visible.

So that became the plan: Take photos during a total solar eclipse that show stars in the vicinity of the Sun. Observe the same star field a few months earlier or later, when there's no spacetime-curving, light-bending Sun in the way. Compare the stellar positions on the two photographs and measure the amount of deflection during the eclipse.

English astronomer Arthur Stanley Eddington was pivotal in turning the plan into reality. News about Einstein's theory of general relativity was slow to reach England in early 1916, because of the ongoing war. But in Leiden, physicists were well aware of the new theory. Willem de Sitter, a brilliant Leiden astronomer and mathematician, wrote about it in the *Monthly Notices of the Royal Astronomical Society.* Eddington happened to be the society's secretary, so he was the first English scientist to learn about Einstein's latest work. He also became one of Einstein's greatest fans and ambassadors.

Earlier German expeditions to measure starlight deflection during the total solar eclipse of August 21, 1914, had been unsuccessful, mainly due to the war. But Eddington believed he could succeed, and he enlisted the help of Frank Dyson, director of the Greenwich Observatory, just east of London, and England's astronomer royal (an honorary position first held by John Flamsteed in 1675).

I can easily imagine the two astronomers discussing Eddington's plan to prove Einstein right. (Mind you, I'm totally making this conversation up.)

"The best opportunity would be the total solar eclipse of May 29, 1919," says Dyson.

"What's so special about that eclipse?" asks Eddington.

"Well, it will be extremely long, as eclipses go. Almost seven minutes. Plenty of time to take photographs. And there's more—during the eclipse, the Sun is in the constellation of Taurus, the Bull. It'll be surrounded by the relatively bright stars of the Hyades, the famous star cluster. Ample opportunity to measure stellar positions."

"So all the news is good? No fine print?"

"Um . . . well," says Dyson, "most of the zone of totality is in the Amazonian rain forest and in the African jungle. There are only two eclipse locations that can be accessed easily: the town of Sobral, in northeastern Brazil, and the isle of Principe, in the Gulf of Guinea."

"Great," replies Eddington. "Let's organize two expeditions, then. If one of those locations is cloudy during the eclipse, we're still fine. If both experience good weather and obtain the same result, we'll have an even more convincing case."

Easier said than done, of course. This was before commercial aviation was widespread, and so people, telescopes, and cameras would all have to go by ship, trips that took weeks. In Brazil, the main telescope failed because of the heat, and Greenwich astronomers Charles Davidson and Andrew Crommelin had to resort to a much smaller instrument. Meanwhile, on Principe, Eddington and clockmaker Edwin Cottingham were plagued by clouds. The only useful photographic plates they brought back home were the ones they managed to expose during the last minute or so of the eclipse.

Chances are you've never seen a total eclipse of the Sun. Most people haven't. They've seen partial ones, where only a certain fraction of the Sun's surface is eclipsed by the Moon, but partial and total eclipses really aren't comparable events. If you have ever witnessed a total eclipse (maybe the one that was visible across a wide swath of the United States on August 21, 2017), I'm sure you'll agree. The sky turns steel blue. Daytime animals become quiet. Darkness sets in, planets and stars become visible, and the silvery white corona of the Sun unfolds around the dark silhouette of the Moon, like a precious gift of nature. It's pure magic.

I've experienced about a dozen total solar eclipses (they're downright addictive—if you've seen one, you want to see more), so I know how Eddington and Cottingham must have felt. On the Caribbean island of Aruba, in February 1998, the sky was overcast nearly all day, almost to the very start of the event. All of us gathered there were freaking out—what if the clouds didn't break in time? (Luckily for us, they did.) A year and a half later, for the eclipse of August 1999, I took my family to Turkey, where the likelihood of clear weather was much greater than in France and Germany. Still, I remember getting extremely nervous the day before the eclipse when a single small cloud appeared above the horizon. And I wasn't even there to prove Einstein right.

Anyway, during the 1919 eclipse, photos were taken, stellar positions were measured, and on Thursday, November 6 of that year, at a joint meeting of the Royal Astronomical Society and the Royal Society of London, Eddington announced the results. Yes, the images of the Hyades stars had all shifted away from the edge of the eclipsed Sun. And yes, the amount of deflection was in good agreement with Einstein's predictions. (Ilse Schneider, one of Einstein's graduate students, later asked him how he would have felt if his prediction had not been confirmed in the 1919 experiment. "Then I would feel sorry for the good Lord," Einstein replied confidently. "The theory is correct anyway.")

The next day, the *London Times* ran a story on the results with the headline "Revolution in Science: New Theory of the Universe." Two days later, on November 9, the *New York Times* put the news on its front page below four of the most memorable headlines I've ever seen.

LIGHTS ALL ASKEW IN THE HEAVENS

MEN OF SCIENCE MORE OR LESS AGOG OVER
RESULTS OF ECLIPSE OBSERVATIONS

EINSTEIN THEORY TRIUMPHS

STARS NOT WHERE THEY SEEMED OR WERE
CALCULATED TO BE, BUT NOBODY NEED WORRY

(I particularly like the "nobody need worry" part: yes, the universe is a mess, but please don't lose any sleep over it.)

Four years after Albert Einstein formulated his theory of general relativity, the world at large finally became aware of it. And people loved it. After the horrors of World War I, which had ended just a year before, they were longing for some good news. What could be better than humankind solving the riddles of the universe? And now that Germany and England were no longer at war, wasn't it great that a German theory had been proven right by English astronomers? Both Einstein and Eddington were committed pacifists, and many people hoped, along with them, that international scientific collaboration would prove to be the antidote for warfare. Almost instantly Einstein became world-famous.

Much later, some scientists started to doubt the accuracy of Eddington's results, and maybe even his scientific honesty. After all, he had been a firm believer in the power of the theory of general relativity from the very start and was extremely eager to prove Einstein right. Might he have been just a bit *too* eager? Might he have thrown away data that weren't in line with Einstein's predictions? Underestimate the measurement errors? Find a result he *wanted* to find?

I don't believe so. Granted, the 1919 photographic plates were of poor quality. Positional uncertainties were pretty large—on the order of a fifth of an arc-second. Present-day astronomers would require results to have a higher degree of statistical significance before they would allow themselves to be convinced of anything. But a 1979 reanalysis of the Sobral and Principe exposures yielded the same result as Eddington found in 1919: the data were compatible with Einstein's theory.

Later eclipse observations have arrived at the same conclusion, at ever higher confidence levels. Moreover, thanks to extremely sensitive space observatories, we don't need solar eclipses anymore to measure the deflection of starlight. The European Gaia mission, launched in December 2013, measures stellar positions to the nearest 1/40,000 of an arc-second. That's the change of direction you'll notice when your friend moves his flashlight by 1 millimeter from a distance of almost 8,500 kilometers instead of 120 meters. Gaia is so sensitive that it measures the light-bending effect of the Sun all over the sky. It even detects the much smaller influence of giant planets such as Jupiter and Saturn.

Finally, astronomers now routinely observe the gravitational lensing effects of large galaxies and clusters of galaxies. Like the Sun, they curve spacetime and bend the paths of light rays from background sources—in this case, extremely remote galaxies. The deflection of starlight is here to stay. Einstein *was* right, after all—at least in this respect.

―――――――

The second testable prediction of general relativity was gravitational redshift. Remember Einstein telling Lorentz that his watch runs a tiny little bit faster on the second floor than in the basement? That's because general relativity predicts that clocks will slow down in strong gravitational fields. Imagine you're in lower Manhattan at ground level. Your sister is at the top of the Freedom Tower, 540 meters up. Now take out your laser pointer. It produces light of one particular wavelength. For green laser pointers, the wavelength is usually 532 nanometers (a nanometer is one-billionth of a meter, so 532 nanometers equals 0.000532 millimeters). Aim the laser pointer at your sister. (Just a reminder that this is a thought experiment—never really aim a laser pointer at someone's face, because it will damage the eye.) What wavelength will she see? Not 532 nanometers, but a slightly longer wavelength, corresponding to a slightly redder color. The reason: time is running faster for her.

Here's how it works. Wavelength is related to frequency, as we saw in Chapter 2. At ground level, your laser pointer produces light with a wavelength of 532 nanometers. This corresponds to a frequency of 563.5 trillion hertz—that's the number of wave crests that pass by in one single second. (Want to do the math yourself? It's easy: divide the speed of light by the wavelength, and you end up with the corresponding frequency.)

At the top of the Freedom Tower, the laser light still has the same velocity. After all, the speed of light is always the same, according to Einstein. But gravity is a bit weaker up there than down at ground level, so time is running a bit faster. Just *before* 563.5 trillion wave crests have passed, one second has already elapsed. In other words, your sister observes a slightly lower frequency, corresponding to a slightly longer wavelength, a slightly lower energy, and a slightly redder color. That's gravitational redshift.

Needless to say, the effect is incredibly small. It's not that the world beneath your feet appears a bit ruddy when you're looking down from a high tower. To give you an idea of the tiny size of this effect, at the summit of Mount Everest time runs faster by about 1 / 30,000 of a second per year as compared to sea level. Your sister would need an extremely sensitive measuring device to detect the very, very slight observed wavelength increase of your laser pointer—it would be less than 0.00000000001 percent.

Robert Pound and Glen Rebka of Harvard University actually built such a measuring device. In 1959, four years after Einstein's death, they carried out the first controlled experiment to measure gravitational redshift. Back then, the Empire State Building was the tallest building in the world. But Pound and Rebka didn't need to take their equipment to New York City. Their device was so incredibly sensitive that the height of Harvard's Jefferson Laboratory—a mere 22.5 meters—was sufficient to detect the effect at a level of one part in 400 trillion.

I'm not going to describe the Pound-Rebka experiment in detail here. It was pretty complicated, using radioactive iron, helium-filled Mylar bags, speaker cones, gamma-ray absorbers, scintillation counters, and what have you. But the take-home message is that the experiment was successful and the results were in excellent agreement with Einstein's theory of general relativity.

So Pound and Rebka confirmed Einstein's prediction that time slows down as gravity increases. With relativity, nothing is absolute anymore—not even the flow of time. And it's not just that the cogs and gears in your watch take longer to wheel around because of some gravitational effect on their mechanics. No, it's really time *itself* that slows down. Each and every physical process takes longer to play itself out in a strong gravitational field.

When I was a teenager, I couldn't really wrap my head around that. I could imagine the hands on my wristwatch moving slower for some reason, but I found it hard to believe that my heart rate would also go down, that my body cells would age at a slower pace, and that I would actually live longer. That sounded like magic or fantasy, not science. Still, that's exactly what's happening.

Then again, my doubt was justified in a certain sense. When time itself slows down in a strong gravitational field (close to a black hole, for instance), every second lasts longer than it normally would. Someone in outer space who has a different frame of reference would indeed notice that my heart is beating more slowly and that I will live longer. But *I* wouldn't be aware of any change at all. There'd be no way for me to notice that my seconds have lengthened. My heart rate would still be a healthy 80 beats per minute. My expected lifetime would still be eighty-plus years or so. The slowing down of time wouldn't be of any advantage to me at all. Even my brain would slow down, so it's not as though I could use the extra time to read more books or learn Chinese.

Anyway, at age fifteen, I had conceptual difficulties with the whole thing, as I'd guess most people do. So I was flabbergasted to read about an exciting experiment, carried out in the fall of 1971, in which physicist Joseph Hafele and astronomer Richard Keating flew around the world in commercial airliners with very special travel companions—atomic clocks—in an attempt to measure the time dilation effect. Total ticket costs: around $8,000, including meals and drinks for the experimenters. Not just exciting, but cheap, too.

First, Hafele and Keating took two atomic clocks on an eastward flight around the world, the same direction as the Earth's rotation. Then they took the clocks (their clocks' official passenger name was "Mr. Clock") on a westward flight around the globe, the opposite direction of the planet's spin. There's an iconic photograph of the two scientists and their instruments filling up an entire row of seats, with a young flight attendant checking her watch, as if it would betray any signs of time dilation. Hafele and Keating aren't with us anymore, but the flight attendant might still be alive and have great stories to tell; alas, I haven't been able to track her down.

Up in the air, where gravity is a teeny bit weaker than on the ground, an atomic clock is expected to run a teeny bit faster. This *gravitational* time dilation had already been convincingly demonstrated by Pound and Rebka in the form of gravitational redshift. But there's also *kinematic* time dilation—an effect predicted by Einstein's 1905 theory of special relativity. To put it simply, the faster you move, the slower your clock runs.

Gravitational time dilation would be comparable for both the eastward and westward flights. After all, both flights took place at very similar altitudes, so the effect of gravity would be the same. But the kinematic time dilation would be different. While on both the eastward and westward flights the plane travels at more or less the same velocity, that's only with respect to the ground beneath. In this case, however, we need to consider the velocities with respect to the center of the Earth. Think of a 3-D coordinate system with the Earth's center at its origin. The planet is rotating in this coordinate system. Earth's surface has a certain rotational velocity at each and every latitude. If you're flying eastward, in the same direction as the Earth's rotation, your velocity with respect to the coordinate system is higher. On a westbound flight, it's lower. And different velocities result in different clock rates.

After landing again in Washington, DC, Hafele and Keating compared their atomic clocks with the one at the United States Naval Observatory. Sure enough, they had gained and lost tens of nanoseconds during their high-velocity flights, in perfect agreement with Einstein's predictions.

An atomic clock works by using fundamental processes at the level of atoms and electrons to measure time. The Hafele-Keating experiment was elegant proof of the fact that each and every physical process in nature is slowed down by time dilation. Physicists may still be ignorant about the true nature of time, but they *do* know that it slows down for observers moving at high velocities or located in strong gravitational fields.

This is good news for astronauts. The International Space Station orbits the Earth at an altitude of a few hundred kilometers. The lower gravity at this altitude means an astronaut's clock is sped up, thanks to gravitational time dilation. But the space station is flying at some eight kilometers per second, and because of this high velocity, the clock is slowed down by *kinematic* time dilation. For orbiting spacecraft, the second effect is larger than the first. The net effect is that once you're on board, you're not growing old as fast as you would on the ground. An astronaut who spends six months on the space station will gain seven milliseconds.

But why is this important? It's all milli- and nanoseconds, parts per trillion, fractions of arc-seconds—how can this have any effect on our daily lives? Isn't this just esoteric entertainment for nerds and geeks who love multiple dimensions, black holes, and weird numbers?

In one sense, the importance of Einstein's theory of general relativity goes beyond anything we might be able to see in our everyday lives, because it tells us about the fundamental properties of the world we live in. Feeling the urge to know, to understand, is an important part of what makes us human.

However, there *are* measurable effects in our daily lives. Not many, but still. For example, your car's GPS navigation wouldn't work properly if technicians didn't take the effects of general relativity into account—you might well end up in a ditch or a river instead of at the restaurant where you have reservations. (This is the striking exception I mentioned in Chapter 1, when I wrote that we can live out our lives perfectly well without understanding anything about general relativity.)

Your navigation system knows where you are. That's how it can guide you from New York to San Francisco, or through the street maze of an unfamiliar town. To calculate your position, the device picks up signals from a handful of satellites that are part of the Global Positioning System (GPS). About thirty of those satellites race around the Earth at an altitude of some 20,000 kilometers. Each of them carries an atomic clock. By comparing the clock signals from three or more GPS satellites, your navigation system works out the distance to each of them. Trigonometry then yields your position—longitude, latitude, and altitude.

But precisely because these satellites are in motion high above the Earth, the GPS clocks experience time dilation, both gravitational and kinematic. If onboard software did not correct for those effects, your calculated position would be off by many meters within an hour. So here's a daily life situation where Einsteinian time slips of nanoseconds *do* matter. Think about that next time you fire up your navigation system.

—————

The Pound-Rebka experiment and the Hafele-Keating experiment are some of the better-known tests of relativity. There have been many more: the Ives-Stilwell experiment, the Kennedy-Thorndike experiment, Rossi-Hall, Frisch-Smith, and the list goes on. (Most are named after two white male experimenters. There are exceptions, however: the Eöt-Wash experiment isn't named after physicists Eöt and Wash, but after Baron Loránd Eötvös de Vásárosnamény and the University of Washington.) I won't describe every experiment here, but the results, whether involving the lifetime of fast-moving muons or the orbital acceleration of the Moon, have confirmed both special and general relativity over and over again, and to ever higher degrees of precision.

Gravity Probe B was the first space-based experiment to test predictions of Einstein's theory of general relativity. The spacecraft's telescope is seen at the upper right; the flat conical structure just above the four solar panels is the dewar that houses the gyroscopes.

So yes, maybe spending $750 million for yet another test could be questioned. Especially if you compare it to the $8,000 spent by Joseph Hafele and Richard Keating to transport themselves and their atomic clocks around the world on jet planes.

Then again, Gravity Probe B was conceived and designed to check on something no one had ever tested before: not time dilation, not gravitational redshift, not deflection of starlight, but geodetic precession and frame dragging. (In case you're wondering, yes, there was also a Gravity Probe A, which was a 1976 experiment to measure gravitational redshift much more precisely than Pound and Rebka had done.)

Geodetic precession is sometimes called de Sitter precession after the Leiden mathematician Willem de Sitter, who first described it back in 1916. (You may remember that de Sitter was also the person whose article brought Einstein's theory of general relativity to England.) It's basically the direct result of spacetime being curved in the neighborhood of a massive body.

Imagine an isolated sphere rotating in empty space. In the absence of any external forces, its rotational axis will always point in the same direction. Now put the rotating sphere in orbit around the Earth. Newton would still expect the rotational axis to maintain its original orientation: if it points in the direction of a distant star, it will keep doing so, orbit after orbit. But Einstein predicts something different. Because of the presence of Earth, spacetime is curved in the planet's vicinity. The sphere's rotational axis indeed keeps a fixed direction in this curved spacetime. But seen from afar, where spacetime is flat again, you'll notice a very slow drift. It may start out pointing in the direction of a distant star, but many orbits later, the alignment will have been lost. That's geodetic precession.

Frame dragging is also easy to visualize. You've probably seen illustrations of spacetime curvature that utilize the image of a bowling ball on a trampoline. The flat surface of the trampoline represents spacetime, and the bowling ball is a massive body like the Sun or a black hole. Just as the bowling ball deforms the surface of the trampoline, massive bodies produce a local curvature in spacetime.

The trampoline metaphor isn't perfect; no single analogy is. But it's useful in the context of frame dragging. Imagine you're standing next to the trampoline. The depression created by the bowling ball is beautifully symmetric. Now press your hand on the top of the bowling ball and make it rotate. The trampoline surface will start to rotate with it. It won't be able to keep up with the rotating bowling ball, though, and so the depression won't be symmetric anymore—all coordinate lines will be twisted in a spiral-like pattern. That's frame dragging.

The "frame" in frame dragging is the so-called rest frame, or the spacetime coordinate system that we're dealing with (the surface of the trampoline). Put a planet (the bowling ball) in your coordinate system, and spacetime gets curved. This curvature produces the geodetic precession described above. Rotate the planet (make the bowling ball spin), and curved spacetime gets dragged along, at least a tiny bit. This produces an additional—and much smaller—precession of the rotational axis of an orbiting body. (This particular type of frame dragging, known as rotational frame dragging, was first predicted by Austrian mathematician Josef Lense and physicist Hans Thirring in 1918, so it's also known as the Lense-Thirring effect.)

At Stanford University, physicists Leonard Schiff and William Fairbank had been toying with the idea of measuring these two effects since 1960. Francis Everitt joined them in 1962, when he was twenty-eight years old. In London, Everitt had been trained as a geologist. But after spending five years in the field of paleomagnetism, he believed that physics would be more interesting, and so he spent an additional two years at the University of Pennsylvania, where he specialized in low-temperature physics.

After he joined the faculty at Stanford, everything came together. The experiment Schiff and Fairbank proposed would make use of ultraprecise gyroscopes—perfect spheres the size of ping-pong balls—that would be magnetized and cooled down close to absolute zero to allow the best possible measurements.

Getting the project going took ages. At first there was little funding—Everitt still wonders how Schiff and Fairbank were able to pay his salary. There was equally little progress. Then NASA got involved, which was both good and bad: the project gained momentum, but the space agency nearly killed the project several times. In the late 1970s, the space shuttle program got started, and NASA decided it wanted to fly Gravity Probe B on the shuttle—the expensive piloted program could use all the scientific justification it could get. Then, in 1986, the *Challenger* exploded, with the loss of seven astronauts. Suddenly no one at NASA wanted to devote resources to a potentially risky physics experiment anymore; even a planned in-flight demonstration on board the shuttle was canceled.

Over the next several years NASA administrators came and went, budgets rose and fell, and visits to Capitol Hill hearings were plentiful. Finally, in the early 1990s, the mission got approved, mainly thanks to project manager Brad Parkinson. Everitt is still convinced that bringing Parkinson on board in the mid-1980s was the most crucial move in Gravity Probe B's erratic history. Not a scientist but an Air Force colonel, inventor, and engineer, Parkinson is credited with having made the Global Positioning System possible, and he knew which strings to pull. Moreover, the Stanford team got the all-important support of Daniel Goldin, NASA's administrator between 1992 and 2001.

Eventually Gravity Probe B was launched from Vandenberg Air Force Base in California on April 20, 2004. Neither Schiff nor Fairbank was alive to see it, and Everitt had just turned seventy. But for him it had been worth the wait.

For about a year, Gravity Probe B's four gyroscopes orbited the Earth in an almost undisturbed free fall, shielded from solar radiation, micrometeorites, and temperature changes by the encapsulating spacecraft. More than 2,400 liters of superfluid liquid helium kept the sensitive science instruments at just 1.8 degrees above absolute zero.

Because of their perfect spherical shape, the gyroscope rotors maintained their orientation with respect to their local frame of reference—the slightly curved spacetime in the vicinity of Earth. Meanwhile, Gravity Probe B's fixed telescope was locked on a distant star in the constellation Pegasus. Geodetic precession and frame dragging would cause the orientation of the gyroscopes to drift very slowly with respect to the satellite. Sensitive superconducting quantum interference devices (SQUIDs) measured a change of alignment in the magnetized rotors of less than 0.0005 arc-seconds.

Very different from boarding a ship to Principe and taking photographs of a solar eclipse, that's for sure. Much more complicated than sending gamma rays from the basement of Harvard's Jefferson Laboratory to the top floor and measuring the tiny wavelength change. And enormously more expensive than flying atomic clocks around the world in commercial airliners. But the experiment offered a unique opportunity to test Einstein's theory of general relativity. If any small deviations from general relativity showed up, the consequences would be enormous.

It took many years to analyze Gravity Probe B's data. The relativistic effects were tiny, and measurement noise was large. Eventually the final results were presented in the spring of 2011, and they were in good agreement with Einstein's predictions. If they hadn't been, the project certainly would have made it to the front page of your local newspaper: "Einstein Was Wrong" would make for a good headline, after all. But no, Einstein was right, again. Geodetic precession: 6.6 arc-seconds per year. Frame dragging: 0.037 arc-seconds per year. Incredibly tiny effects, but almost exactly the predicted values. Never before had general relativity been tested and confirmed to such a high level of precision. So don't try telling Francis Everitt that the project's $750 million cost was not worth it.

So are we finally done with trying to prove Einstein's theories right or wrong?

No way.

General relativity in its current form may very well not be the final word on the nature of space, time, and gravity. The reason: the theory is completely incompatible with quantum mechanics, that other momentous pillar of twentieth-century physics. It's a problem I'll return to in Chapter 12. Sooner or later, scientists are bound to come up with experimental results that do not fully confirm the predictions of one or the other of those two theories, just as the odd behavior of Mercury's orbit didn't comply with Newton's theory. These would be the scientific equivalent of small clouds on the horizon—innocent at first, but with the potential of growing into massive thunderstorms. The result would be hints pointing the way to newer and better theories.

It's not surprising, then, that the first direct detection of gravitational waves, in September 2015, was hailed as one of the most important scientific breakthroughs in decades. Here was a century-old prediction by Albert Einstein that had *never* been directly confirmed. It was also a novel way of studying the most enigmatic objects in the universe—black holes.

Might this new tool provide us with the key to unlocking the mysteries of spacetime?

((⟨ 4 ⟩))

Wave Talk and Bar Fights

Philip Morrison had nothing but his cane to wield.

It was Monday, June 10, 1974. Dozens of physicists had gathered at the Massachusetts Institute of Technology (MIT) for the Fifth Cambridge Conference on Relativity. Invited lectures, contributed talks, poster presentations, Q&A sessions—nothing special. A science meeting like any other.

Until the topic of gravitational waves came up. Two prominent conference attendees, Joe Weber and Dick Garwin, had a discussion that turned into an argument. Then they started to shout and call each other names. Next they rose to their feet and approached each other in front of the audience, eyes furious, jaws tight, fists clenched. *What the heck?*

Morrison, a polio-ridden physics professor at MIT, was the session moderator. His "Gentlemen, gentlemen" didn't help. Any moment now, Weber and Garwin might end up in something that resembled a bar fight. What could Morrison do? Like a wizard wielding his magic wand, he raised his cane to separate the two warriors. It worked. No blood was spilled.

So what was it all about? Simply put, Joe Weber claimed to have detected gravitational waves. Dick Garwin didn't believe him, and he had very good reasons to be skeptical. In fact, hardly anyone believed

Weber's claim. Back then, some physicists even doubted that gravitational waves existed at all. No wonder people became emotional.

———

Confusion about gravitational waves dates back to 1916 and to Albert Einstein himself. The reason? Not every prediction of general relativity is as clear-cut as one might hope. Sure, Mercury's perihelion should precess faster than Newton would expect. Starlight should be bent by the curvature of spacetime. Time should slow down in strong gravitational fields. Those were the easy predictions. But others were less obvious, and the existence of gravitational waves is one of those. At least it was to Einstein.

In mathematical terms, the field equations of general relativity were more or less similar to Maxwell's equations of electrodynamics. In the 1860s, Scottish physicist James Clerk Maxwell first suggested that electricity and magnetism are really two sides of a single coin. He also proposed that light is an electromagnetic wave phenomenon. A century and a half later, his equations are still famous enough to have appeared on T-shirts (although physics students are probably the only ones to wear them). The same is true for Einstein's field equations.

But how similar is similar?

Maxwell's theory of electrodynamics is pretty straightforward. Take an electric charge, accelerate it, and it will produce an electromagnetic wave. We experience the result all around us in the form of light, radio waves, and so on. So, naively, you might expect the same to be true for general relativity: take a "gravitational charge" (a massive object), accelerate it, and it will produce a gravitational wave. Sounds logical. It was certainly something Einstein considered in late 1915, after he had worked out the final version of his field equations.

But there's a big difference between electromagnetism and gravity. Electric and magnetic charges can be both positive and negative. They can attract or repel each other. Mass, however, is always positive. There's no such thing as negative mass. As a result, gravity is always attractive and never repulsive.

In early 1916, this led Einstein to conclude that "there are no gravitational waves analogous to light waves." That's what he wrote in a letter to German mathematician Karl Schwarzschild. His complicated argument had to do with scalars, tensor densities, dipoles, and unimodular coordinate systems (you don't need to know what those are; I just mention them to stress that general relativity is not a walk in the park).

Later that year, Einstein reversed his opinion completely after Willem de Sitter in Leiden suggested using a different coordinate system for the calculations. It made a huge difference. Yes, Einstein concluded, gravitational waves *do* exist. And they propagate at the speed of light—just like Maxwell's electromagnetic waves. In June, Einstein presented his newest results to the Prussian Academy of Sciences in Berlin. "Approximate Integration of the Field Equations of Gravitation" may not sound too exciting, but it's a landmark paper. It's the first publication about gravitational waves ever.

And it's wrong.

In the fall of 1917, Finnish physicist Gunnar Nordström pointed out an important error in Einstein's work. (In case you're curious, it had to do with the derivation of a pseudotensor.) Because of the error, the formulas for gravitational waves that Einstein came up with in 1916 were off the mark. So maybe his January 1918 paper, simply titled "On Gravitational Waves," should earn the landmark label instead. "I have to return to the subject matter," Einstein wrote in the very first paragraph, "since my former presentation is not sufficiently transparent and, furthermore, is marred by a regrettable error in calculation." It's always good to be honest about your mistakes, especially in science.

Not that the 1918 paper convinced everyone. A particularly vocal critic of gravitational waves was Arthur Stanley Eddington, of all people. Yes, one of Einstein's greatest fans and one of the first popularizers of general relativity, and himself an eminent astrophysicist.

Eddington believed gravitational waves were a mathematical quirk of the theory, without any physical meaning. He also took issue with Einstein's conclusion that they would travel at light speed. In

1922, he famously stated that "gravitational waves propagate at the speed of thought"—a sly way of saying that they're a figment of the imagination.

In the 1920s and early 1930s, no one paid too much attention to the notion of gravitational waves. After all, if they existed, they would be far too small to detect. It seemed impossible that scientists would ever be able to confirm or refute the prediction. Most people completely forgot about them.

It wasn't until 1936 that Einstein returned to the topic. By then, he was living in the United States and held a position at Princeton's Institute for Advanced Study. Great place, great people, great minds. He particularly enjoyed working with Nathan Rosen, who was young enough to be Einstein's son. Together, they worked on general relativity, quantum mechanics, wormholes—and gravitational waves. And Einstein and Rosen came to the surprising conclusion that the waves did not exist after all; apparently Eddington had been right all along. Before long, they submitted a paper to the leading professional physics journal of the time, *Physical Review.* The paper's title: "Do Gravitational Waves Exist?" The message: no, they don't, and here's why not.

Einstein and Rosen were wrong, of course—just ask the thousand-plus scientists of the LIGO Scientific Collaboration and the Virgo Collaboration, who announced the first detection of gravitational waves in February 2016. So in the end, it's a good thing that the paper was never published. John Tate, the editor of *Physical Review,* had sent the manuscript to a referee, who advised against publication. "As far as I can see," the referee wrote, "the . . . objections of Einstein and Rosen [against the existence of gravitational waves] do not exist."

Having science papers refereed by anonymous peers is common practice these days, especially in the physical sciences. But back then, it was a pretty new way of going about things, even for *Physical Review.* Einstein wasn't familiar with it at all. In Europe, journals simply published what scientists sent them. He was furious about the rejection, and would never publish in *Physical Review* again. Instead, he submitted the paper to the much smaller *Journal of the Franklin Institute* in Philadelphia, which didn't utilize peer review, and the paper was readily accepted for publication.

But in the fall of 1936, things changed. Nathan Rosen left to take a position in the Soviet Union, and Polish physicist Leopold Infeld became Einstein's new assistant. Cosmologist Howard Robertson explained to Infeld where Einstein and Rosen had erred. (Robertson, in fact, had been the referee for the *Physical Review* paper.) By the time Infeld told his boss that there was a problem, Einstein himself had also discovered the error. Even Nathan Rosen, in distant Kiev, came across the problem, which was of an esoteric mathematical nature.

So in the end, the paper that appeared in the *Journal of the Franklin Institute* in January 1937 was a much revised version. Einstein also changed the title. Just like his 1918 publication (also a corrected version of an earlier paper), it was now called simply "On Gravitational Waves." Its main message was as follows: we can't prove that the elusive waves don't exist, but we're not sure they do either.

By now, the theory of general relativity was nearly twenty-five years old. But scientists continued to disagree about the existence of something the theory could be seen to predict. This remained the state of affairs for the next twenty years. When Einstein died in 1955, the physical reality of gravitational waves was still very much debated, their properties still very much unknown. For instance, less than three months after Einstein's death, Rosen argued that gravitational waves could not carry any energy—another way of saying that they couldn't have real physical existence. But a year and a half later, opinions started to change, especially after theoretical physicists Felix Pirani and Richard Feynman and cosmologist Hermann Bondi proved that they *would* carry energy. Gravitational waves entered the realm of real physical phenomena. Now the only remaining problem was to catch them.

———

Before we go on, we need to have a really good mental picture of what gravitational waves are. I'm sure you've heard the phrase "ripples in the fabric of spacetime." You may have also seen animations of black holes merging, emitting spiral-shaped undulations in a two-dimensional plane. Let me try yet another way of explaining these intriguing Einstein waves. ("Einstein waves" is not an official term. But I like the ring of it, and I'll take the liberty of using it as a synonym for "gravitational waves" every now and then.)

First, the most important thing to realize is that it's not that something is "waving" or "rippling" *in* space, like water waves, sound waves, or even light waves. No, this is about spacetime *itself*. To visualize this, let's first consider one-dimensional "space"—a straight line. Think of a taut jump rope. It's possible to set up waves in the rope by moving one end up and down regularly. But if we want to understand Einstein waves, this image is altogether wrong. Remember, we're talking waves of (and in) space itself. If space is one-dimensional, we should imagine ripples *in* that one single dimension.

A plastic jump rope is somewhat elastic—it can be stretched a little bit in one place and compressed slightly in another, so the total length of the rope doesn't change. It also remains a one-dimensional straight line. But it's possible for longitudinal waves to propagate through the rope. Suppose there are tick marks on the rope at one-millimeter intervals. If such a longitudinal wave propagates through the rope, you would first see the tick marks move away from each other and then move closer again. This is a neat way of visualizing a one-dimensional gravitational wave. Space is alternately stretched and squeezed.

Now move on to a two-dimensional space, like a sheet of graph paper. Here it's exactly the same. A gravitational wave in two-dimensional space should not be depicted as an undulating sheet of graph paper, as is often done. Instead, we should try to visualize a propagating ripple *in* the two-dimensional plane. The result is that squares of the graph paper are stretched out at some place and compressed at other places. (Or, more precisely, at one time a particular square grows in a certain direction; the next moment it shrinks.) Perpendicular to the direction of the wave, space is alternately stretched and squeezed. It's as if regions of higher and lower "space density" are propagating through the plane.

So what about Einstein waves in three-dimensional space? Well, there's no need to suddenly imagine funny ripples in a hypothetical fourth dimension. It's just the propagation of "space density" ripples. Imagine 3-D graph paper made up of cubes, and notice how their sides grow longer and shorter perpendicular to the direction of the wave as it passes through.

Waves in 3-D space are, of course, three-dimensional. Popular images and movies that represent them two-dimensionally wrongly give the impression that two orbiting black holes emit gravitational waves only in a horizontal plane. Instead, the waves spread out in all directions. They may be stronger in one direction than in the other, but you shouldn't be lured into thinking that they're only emitted in the orbital plane.

So that's the proper way to visualize Einstein waves. It's really not very different from density ripples propagating through a bowl of Jell-O that's being tapped on, where the Jell-O represents empty space.

Depending on the source of gravitational waves, they can have very different frequencies and amplitudes. (If you've forgotten about frequency, wavelength, amplitude, and velocity of wave phenomena, just turn back to Chapter 2.) Imagine two orbiting black holes really close to each other. Let's assume they complete 100 orbits per second (yes, that's a very realistic number). From Einstein's theory, it follows that they emit gravitational waves at a frequency of 200 hertz—for an observer at some distance, 200 "wave crests" pass by each second. Since gravitational waves travel at the speed of light (300,000 kilometers per second), the corresponding wavelength is 1,500 kilometers.

What about the amplitude? The amplitude of a gravitational wave is a measure of its strength. It tells you to what degree spacetime is being stretched and compressed. There are two important things to realize here. First, amplitude decreases with distance. In the vicinity of the orbiting black holes, the spacetime ripples are stronger than they are when they've moved farther away. In fact, the amplitude is inversely proportional to the distance. Simply put, when the waves have moved five times farther out, they have become five times weaker.

(This may sound strange. After all, the force of gravity or the brightness of a light source diminishes with the *square* of the distance. Move a pair of planets five times farther apart, and their mutual gravitational attraction becomes twenty-five times weaker. Move a star ten times farther away, and it becomes a hundred times fainter. But in those cases, we're considering the energy in a gravitational field or in a light wave. With Einstein waves, we're talking about the amplitude, which really is inversely proportional to distance.)

The other thing to realize is that the amplitude of gravitational waves is unbelievably small. I compared empty space to a bowl of Jell-O. But it would be better to compare it to a block of concrete. If I gently tap a bowl of Jell-O, it starts to quiver all over. But even if I slam a block of concrete with a sledgehammer, you'll hardly notice ripples propagating through the stuff. That's simply because concrete is much stiffer than Jell-O. Similarly, spacetime is incredibly stiff. It's not easy to flex, bend, stretch, or compress. You need *a lot* of energy to produce even the tiniest ripples.

So that's the gravitational wave signal of our two orbiting black holes. Velocity: speed of light. Frequency: 200 hertz. Corresponding wavelength: 1,500 kilometers. Amplitude: inversely proportional to the distance between the observer and the black hole pair, but in any case incredibly small.

What if the black holes were much more massive? Well, if they still orbited each other a hundred times per second, the frequency (and the wavelength, of course) would be exactly the same. The amplitude would be larger, though, because of the higher masses.

However, the amplitude is also dependent on the acceleration of the orbiting black holes. Move them farther in toward each other, where they whirl around at higher velocities, and the amplitude will increase even more. The frequency will increase, too: with a smaller distance between them, the black holes will have a smaller orbital period. So if two black holes spiral in toward each other, the gravitational wave signal will show an increase in both amplitude and frequency. That's exactly what the LIGO detectors saw in September 2015, when they registered their first Einstein waves.

There's much more to tell, but I'll keep that for later chapters. It's time to move on to more exciting stories, such as that of scientists who are about to assault each other in front of a conference audience.

———

Joseph Weber knew all about fighting. During World War II, he was a US Navy lieutenant commander. In May 1942, he barely escaped the sinking USS *Lexington* after the Japanese turned it into a tangled mass of burning steel. Joe was about to turn thirty-three—he was born twelve days before Arthur Eddington cursed the clouds on the isle of Principe.

After the war, Weber worked as an electrical engineer at the University of Maryland in College Park, just northeast of Washington, DC. He obtained a PhD in microwave spectroscopy and even worked out the fundamentals of lasers and masers—the first moves toward developments that would earn others the Nobel Prize in Physics in 1964.

Weber's interest in relativity and gravity surfaced in the mid-1950s, when he spent a sabbatical year with physics guru John Archibald Wheeler both in Princeton and in Leiden. Curved spacetime, black holes, time dilation, gravitational waves—great stuff! He set out to learn everything he could about it, and in 1961, he published a small book titled *General Relativity and Gravitational Waves*.

By then, though, he had already published the idea that was to make him famous—or infamous, as some people might say. Joe Weber decided to open the hunt for Einstein waves. There'd been so much theoretical talk about them over the years; now it was time to roll up his sleeves, build instruments, and do experiments to try to catch the things.

The plan was simple: just measure the minute, periodic change in size of something here on Earth. After all, a passing gravitational wave stretches and squeezes empty space *and everything in it*. A block of concrete will actually grow and shrink a tiny little bit in response to passing gravitational waves. However, the change in size would be infinitesimally small, thus extremely hard to measure. Moreover, using a ruler wouldn't work because the ruler, too, would grow and shrink.

But Weber had a solution: natural frequencies.

Most objects have a certain natural frequency at which vibrations tend to resonate and amplify themselves. Ask the older inhabitants of Tacoma, Washington, just south of Seattle. They will remember the dramatic collapse, in November 1940, of a huge, newly built suspension bridge that connected the city to the Kitsap Peninsula. Apparently the bridge's natural frequency matched the prevailing frequency of strong gusts of wind across the Tacoma Narrows. The whole structure started to resonate, vibrate, and twist, until it snapped. Footage of the bridge's collapse—look for it on YouTube—is jaw-dropping.

So here's Weber's plan. Use a big aluminum cylinder as your detector. Have it precisely machined so it has a very specific natural frequency. Suspend it on a steel wire to isolate it from environmental vibrations. Place everything in a vacuum tank for the same reason. Attach piezoelectric sensors to the cylinder. Then wait.

If gravitational waves exist, they will have a wide range of frequencies. Supernova explosions, colliding stars, orbiting black holes—each astrophysical event has its own signature frequency. Upon arriving at the Earth, they will cause minute vibrations in the aluminum cylinder. The hope is that some Einstein waves will have the same frequency as the natural frequency of the aluminum cylinder, making it start to resonate. When that happens, its vibrations will become stronger, maybe even measurable. Moreover, seconds after the wave has passed, the bar will continue to vibrate, like a tuning fork that has been struck. The piezoelectric sensors will register the rapid stretching and squeezing of the bar, turning the tiny quivers into an electric signal.

In the early 1960s, Weber and his postdoc Bob Forward built and deployed the first "resonant gravitational-wave detectors," as they were called. Or "resonant bar antennas." Or just "Weber bars." Sure enough, they picked up tiny signals every now and then—something that seemed to stand out from the ever-present vibrational background noise. A supernova in a remote galaxy? Colliding neutron stars in our cosmic backyard? Some unknown energetic process in the core of the Milky Way? Who knows?

(When I first learned of Weber's collaboration with Robert L. Forward, I thought, "That's funny, the guy has the same name as the author of *Dragon's Egg*"—a 1980 science fiction novel about life on the surface of a neutron star. It turned out to be the same person. He left the University of Maryland in 1962.)

Weber's experiments started to really draw attention when, in 1968, he used two identical detectors, one at the Maryland campus in College Park and the other about a thousand kilometers farther west at the Argonne National Laboratory near Chicago. The intention was to eliminate false positives. A passing truck on Baltimore Avenue might set off a vibration in the College Park bar but not in Chicago. However, gravitational waves from a supernova explosion or a stellar collision should register at both sites simultaneously—or at least within a fraction of a second, given the speed of the waves and depending on the direction of origin.

The two aluminum bar antennas each had a length of a meter and a half, measured some 65 centimeters in diameter, and weighed in at 1,400 kilograms. Their natural frequency was 1,660 hertz—in the right ballpark for Einstein waves from two colliding neutron stars. (We'll turn to neutron stars in Chapter 5.) So now it was just a matter of waiting for two signals to be detected at the same time, a phenomenon called a "coincidence."

Weber didn't have to wait long. Between December 30, 1968, and March 21, 1969, no fewer than seventeen coincidences were detected. Surely that couldn't be due to chance. In early June, his first public announcement of these results at a relativity conference in Cincinnati, Ohio, was greeted with applause. Shortly thereafter, on June 16, *Physical Review Letters* published his paper "Evidence for Discovery of Gravitational Radiation" (gravitational radiation is just a now-obsolete synonym for gravitational waves).

Excitement soon turned into doubt. First, astrophysicists wondered about the sheer number of events. Given the sensitivity of Weber's bar antennas, waves produced by the collision of neutron stars would have

Joe Weber checks the display of his gravitational wave experiment. In the background is a vacuum tank housing one of his aluminum resonant bar detectors.

had to have been produced within a few hundred light-years of Earth. In such a small volume of space, seventeen stellar collisions in three months was outright impossible. But if the waves came from much farther away—some unknown energetic process in the core of the Milky Way, for instance—the energies involved would have to be un-believably large.

Experimentalists became skeptical, too. To be considered valid by the scientific community, experimental results have to be replicable. But at Moscow State University, Vladimir Braginsky couldn't repeat Weber's results. Anthony Tyson at Bell Telephone Laboratories in Hol-mdel, New Jersey, found nothing. For David Douglass at the University of Rochester, it was a negative. Ron Drever in Glasgow found zilch. Meanwhile, Weber kept reporting new detections from his Maryland "gravitational wave observatory."

Tony Tyson still recalls his discussion with Al Clogston, who di-rected the Physics Research Laboratory at Bell Labs. When Tyson told him he planned an experiment to check Weber's results, Clogston was less than enthusiastic, mostly because there seemed to be no upside for Tyson and the lab: if Weber was proven wrong, nothing would be gained, but if he was proven right, it would be Weber who got the Nobel, not Tyson, so why bother? Still, Tyson began building very sensitive resonant detectors, more or less secretly. He teamed up with Dave Douglass, and they even started to cooperate with Weber in 1971, comparing the readings at Holmdel with those from Rochester, sharing data with Maryland, improving the equipment's sensitivity, and writing better analysis software.

By late 1972, Tyson was convinced that Weber was seeing things that weren't there. Weber was a brilliant thinker and a clever engineer, but he was sloppy when it came to data analysis and statistics. He never published the algorithms that he used to define and identify co-incidences in the readings from his various bar antennas. If you keep changing the criteria you use, you'll always be able to find as many "coincidences" as you want.

Weber made silly errors, too. He claimed he saw signals from the heart of the Milky Way because detections appeared to occur preferentially when the Milky Way's center was high up in the sky, and Einstein waves coming from above would produce stronger signals in the bars than when they came in horizontally. That's correct, but Tyson had to remind him that the Earth is transparent to gravitational waves. As a result, signals should have been equally strong when the Milky Way reached its deepest point *below* the horizon, though Weber did not report any.

Later, Weber claimed to have found coincidences between his own measurements and the Holmdel and Rochester data—signals that, admittedly, barely stood out against the noise but appeared to show up at exactly the same time. But then Tyson and Douglass discovered that Weber was using Eastern Daylight Time while *they* worked with Universal Time, which was four hours off. Very embarrassing.

It was a roller-coaster period for Joe Weber. He worked long days in the lab alone and was constantly dealing with criticism of his work. In the summer of 1971, his wife died of a heart attack. But Weber was stubborn and wouldn't give up. In March 1972, at age fifty-two, he married again, this time to Virginia Trimble, a twenty-eight-year-old astronomer from California, and he started taking dancing lessons.

However, the bar fights continued. By 1974, many Weber bar experiments were running all over the world. Tyson and Douglass were now operating four-ton instruments with low-temperature electronics to fight the ever-present measurement noise. Still, they saw nothing. In Munich, Germany, at the Max Planck Institute for Astrophysics, Heinz Billing, Albrecht Rüdiger, and Ronald Schilling built a large bar detector, as did Guido Pizzella and Karl Maischberger in Frascati, Italy. Zero results. And then there was Richard Garwin's small instrument at IBM's Thomas J. Watson Research Center in Yorktown Heights, New York. It weighed only 120 kilograms and so would be able to detect only the most powerful gravitational waves, but even so, it found nothing.

Dick Garwin was not someone to toy with. In 1952, at age twenty-four, he had worked for Edward Teller on the hydrogen bomb. He was a brilliant physicist and a highly esteemed government consultant on matters of national security, and he had served two terms on the President's Science Advisory Committee. Moreover, he had a better grasp of how to manage data than Weber did.

Tony Tyson had already clashed with Joe Weber over gravitational waves at a large conference in New York City in December 1972. (That was the sixth Texas Symposium on Relativistic Astrophysics; New York City is of course not in Texas, but the first conference in the series was, and the name stuck.) But that had been a more or less polite scientific argument. Despite their disagreement on the data, Tyson and Weber respected each other. Many years later, they even became friends—sort of.

With Garwin at the Cambridge conference in June 1974, on the other hand, things were different. Maybe it was because Weber was tired of defending himself. Or maybe he knew deep down that something was wrong. We'll never know. At any rate, he experienced Garwin's critique as a personal attack, and he was ready to fight back—until Phil Morrison intervened.

Looking back at the episode more than forty years later, Virginia Trimble still feels sorry for her late husband. "They voted him off the island," she told me, using a reference to the popular TV series *Survivor*. "You don't know the meaning of the word *controversial* unless you've been married to Joe Weber for twenty-eight years. [The physics community] was a tribe. And in the end, Garwin was the most persistent of them all. To Joe, he was evil incarnate."

Trimble, who became an eminent astrophysicist and historian of astronomy, never got involved in the debate over detecting gravitational waves with the bar detectors. And her career didn't suffer because of her relationship with Weber. After his death, she sold their house in Chevy Chase and used the proceeds to establish the American Astronomical Society's Joseph Weber Award for Astronomical Instrumentation. Since 2002, it has been awarded to individuals who had an attitude like Weber's: build the best detector you know how to build, and operate it until you understand what you are seeing.

After the Cambridge confrontation, the argument between Weber and Garwin continued—not at conferences, but in the letters section of *Physics Today.* In June 1975, Princeton physicist Freeman Dyson sent Weber a letter suggesting that he surrender. "A great man is not afraid to admit publicly that he has made a mistake and has changed his mind," Dyson wrote. "You are strong enough to admit that you are wrong. If you do this, your enemies will rejoice but your friends will rejoice even more." But Weber refused to concede.

By that time, most scientists were convinced that Weber's claims were unsubstantiated—not because there's something wrong with the bar detector technology as such but apparently because gravitational waves are too weak to be measured this way. Numerous resonant detectors have been built and operated in many places since the mid-1970s, employing different sizes, different shapes, different materials, and different masses. The best were extremely sensitive, superbly isolated from vibrational noise (like that from passing trucks), cryogenic (cooled down close to absolute zero, −273 degrees Celsius), and equipped with superconducting quantum interference devices to detect the smallest possible signal. And while sometimes it looked like one of them might have found something, the data were never convincing enough to the critics, and most of the detectors were eventually decommissioned. Meanwhile, Weber lost his National Science Foundation funding in the late 1980s. Partly using his own money, he kept his bars running until his death in September 2000. Some of his original hardware is still sitting in small, garage-like buildings at the University of Maryland campus, doing nothing but collecting dust.

———

It's a sad story, and you can't help feeling sorry for Joe Weber. But this is often the fate of a pioneer. Opening up a new field of research is the hardest possible thing. If what you're after were easy, everyone would be doing it already. If you're the first, chances are you won't succeed, for whatever reason.

One astronomer who would later share a Nobel Prize for work related to Einstein waves was not attending the Fifth Cambridge Conference on Relativity in June 1974; he wasn't even aware of the Weber bar controversy. At age twenty-three, Russell Hulse was observing pulsars at the Arecibo radio observatory in Puerto Rico as part of his PhD thesis work. That summer, he made the discovery that led to the first (indirect) proof of the existence of gravitational waves.

Before we delve into that story, however, you need to understand what neutron stars are. So buckle up for a crash course in astrophysics.

(((5)))

The Lives of Stars

Remember Carl Sagan, planetary scientist, astronomy popularizer, and presenter of the 1980 PBS TV series *Cosmos: A Personal Voyage*? If that series was broadcast before you were born, Google it—it's very much worth watching.

Episode 9 starts with slow-motion close-ups of the making of an apple pie accompanied by classical music. Then a formally dressed server bears the pie on a silver platter through a refectory at Cambridge University to Sagan, who is seated at an elegantly set table in the front of the large room. After the pie is placed in front of him, Sagan looks into the camera and says, "If you wish to make an apple pie from scratch, you must first invent the universe."

It's true, of course. Without the big bang, there would be no galaxies, stars, or planets, let alone apple pies. Everything around you has its own particular history. Chairs, cats, or car keys—to really understand them, you must know where they came from.

The same holds for neutron stars. To paraphrase Sagan, if you want to know what a neutron star is, you must first understand stellar evolution. After all, a neutron star is the corpse of a star. We need a solid knowledge of neutron stars for our story on gravitational waves, so I'm going to give you an introductory course on the lives of stars. And eventually I'll come back around to Sagan's apple pie.

Stars are important. For one, they provide the energy for living organisms. Life on Earth, for instance, is fully dependent on energy from the Sun. Without the Sun's energy, the Earth would be a dark, frigid rock. Nothing could survive.

If we're that dependent on the Sun, we'd better understand how it works and what it's made of. Where does all the energy come from? How long will it last? What happens if the Sun dies? Astronomers didn't know the answers to those questions until less than a century ago. After all, there's no way to study the Sun in a laboratory or to examine a sample of solar material under the microscope.

Little wonder, then, that at the beginning of the Industrial Revolution, some people believed the Sun was made of coal, the new miracle energy source. Heap up enough of the black stuff, they thought, and it might start to shine. Nineteenth-century scientists were a bit more realistic, thinking the Sun might be slowly shrinking, or perhaps that it was constantly bombarded by meteorites. Both processes would release energy.

They were wrong. The Sun isn't shrinking; in fact, it's growing in size, albeit at an imperceptibly slow rate. And yes, meteorites and even comets plunge into the Sun all the time, but the impact rate is much too low to explain the Sun's heat and light. As for coal, if the Sun were a coal-fired power plant, it would last for a mere 6,000 years or so. While that might fit in with the worldview of some creationists, it's really less than two-millionths of the age of terrestrial life.

Enter Cecilia Payne. At the age of nineteen, Payne became interested in astronomy when she heard about Arthur Eddington's eclipse expedition that confirmed Einstein's general theory of relativity—you read about this in Chapter 3. Four years later, she left England for a fellowship at Harvard College Observatory, earning the first PhD in astronomy from Radcliffe College. In her 1925 thesis, she showed that the Sun is mostly made of hydrogen, the simplest element in nature. And, since the same must be true for other stars, Payne essentially discovered the composition of the universe. It's pretty embarrassing, then, that most people have never heard of her.

We now know that the Sun is 71 percent hydrogen, 27 percent helium (nature's second-simplest element), and just 2 percent heavier elements. In fact, the Sun is nothing more than a huge ball of hot gas. Maybe *huge* is not the right word; *humongous* would be better. It's a whopping 1.4 million kilometers across—over a hundred times the diameter of Earth. If the Sun were the size of a beach ball, Earth would be as small as a marble in comparison—just try to visualize that. And if the Sun also were as *hollow* as a beach ball, it could contain over 1.3 million of those Earth-sized blue marbles. Quite impressive.

So how does a humongous ball of hydrogen and helium produce a constant flow of energy? Easy, nuclear fusion. Well, maybe it's not *that* easy—it took American physicist Hans Bethe until the late 1930s to work out the details. But if we forget about the details, it's really straightforward. In the core of the Sun, the gas is strongly compressed by the weight of the overlying layers; the density there is thirteen times the density of lead. Under these extreme circumstances, atomic nuclei start to fuse together—there's your nuclear fusion. And if you've ever seen footage of the test of the first American hydrogen bomb, which was developed in the early 1950s, you know that nuclear fusion releases energy. Lots of energy.

Let's do a thought experiment. Let's imagine that we can switch on the nuclear fusion reactions in the Sun's core for just one second and then switch them off again. What would happen during that one particular second? (Flabbergast alert: what follows may be hard to imagine, but it's true nevertheless.)

In just one second, 570 million tons of hydrogen gas take part in nuclear fusion reactions. That's about the mass of a cube of concrete that's over 600 meters on each side. If you *really* fancy large numbers, it's some 3.4×10^{38} hydrogen nuclei. Yes, that's in one second. These lightweight hydrogen nuclei (in fact, single protons) are fused into more massive nuclei of helium atoms. A helium nucleus is about four times the mass of a proton, so for every four hydrogen nuclei that go into the nuclear fusion black box, one helium nucleus comes out. (That's still a lot, as you'll realize when you divide 3.4×10^{38} by 4— you end up with 8.5×10^{37}.)

By the way, I just introduced the scientific notation of large numbers. In case you're unfamiliar with it, it has to do with shifting decimal places: 3.4×10^{38} means take the number 3.4 and shift the decimal place thirty-eight positions to the right, adding zeros as you go. You'll end up with 340,000,000,000,000,000,000,000,000,000,000,000,000. Likewise, 3.4×10^{-20} means you have to shift the decimal point twenty positions to the left, which yields 0.000,000,000,000,000,000,034. Astronomy is the science of large numbers; astronomy books that didn't use scientific notation every now and then would just use up too many trees.

So every single second, a huge number of protons (hydrogen nuclei) fuse into helium nuclei. Now comes the tricky part. I just said that a helium nucleus is *about* four times the mass of a proton. In fact, it's a teeny-weeny bit shy of that. In goes 570 million tons of hydrogen, out comes a "mere" 566 million tons of helium—0.7 percent less. So what has happened to the remaining 4 million tons? You may have guessed already: it has been converted into energy. $E = mc^2$: Einstein all over again.

So during our one-second thought experiment, the Sun has lost 4 million tons of mass. Now *that's* what I'd call losing weight effectively. In case you're wondering how there can be any Sun left at all, just do the calculations. If the loss of mass has been constant over the Sun's lifetime of 4.6 billion years (145 quadrillion seconds), the Sun is now 6×10^{23} tons less massive than when it was born. But that's only 0.03 percent of its total mass of 2×10^{27} tons—nothing that spectacular. In fact, I take back what I just said about losing weight effectively: for a person weighing 100 kilograms, 0.03 percent would amount to just 30 grams.

Not all of the lost mass goes into energy. The fusion of four hydrogen nuclei into one helium nucleus also produces two positrons and two neutrinos. But together, the two positrons weigh less than

0.1 percent of a hydrogen nucleus, and neutrinos are effectively massless. For now, we may as well forget about those particles (we'll get back to the neutrinos, though). So the upshot is that the Sun converts 4 million tons of mass into pure energy each and every second. That's a lot of energy: 400 quadrillion gigajoules, or about a million times the annual energy consumption of all of humanity, in a single second. If only we could harness one second's worth of the Sun's nuclear power, there would be no energy crisis until the year AD 1,002,000.

Our one-second thought experiment is now over, and the nuclear fusion reactions have miraculously shut down. What happens to the energy? It has been produced in the form of energetic gamma rays, but those are pretty much locked up in the Sun's interior. Remember, the density in the core is very high, and the 15-million-degree gas is almost completely opaque. Gamma-ray photons can't travel far. They interact strongly with the gas particles. As a result, the energy that was released in our single second is being absorbed, reemitted, and scattered in random directions within the Sun, only to be absorbed and reemitted again and again. Hardly a way to make rapid progress.

In a perfect vacuum, light travels at 300,000 kilometers per second. Naively, you might expect the radiation from the Sun's interior to reach the solar surface in just over two seconds—after all, the trip is a mere 700,000 kilometers. But in reality, it takes about 100,000 years, thanks to the Sun's opaqueness. So the 400 quadrillion gigajoules' worth of nuclear energy produced during the one second of our thought experiment won't reach the surface of the Sun until 100,000 years from now. From there, it takes only eight minutes and twenty seconds for that light to travel through the near vacuum of interplanetary space and reach Earth.

Of course, this also means that the energy we receive from the Sun today was produced about a 100,000 years ago. In a sense, we're basking in solar energy that dates back to the time of primitive *Homo sapiens*. And if for some reason the nuclear fusion reactions in the Sun's interior really did shut down at this very moment, we'd be safe for another 5,000 generations or so.

So now we know what the Sun is composed of and how it produces energy. The same story holds for all the stars in the night sky. They're nuclear power plants of hydrogen and helium, pumping out incredible amounts of energy every single second. But to learn about neutron stars, since that's what we're after, we also need to know about the birth and death of stars.

———

The stars haven't always been around, and they won't last forever. Stars are born, they live out their lives, and then they die. (A star is not a living being, of course, but the metaphor is too useful to abandon—even professional astronomers talk about the birth and death of stars.) Our own Sun is a middle-aged star: it was born some 4.6 billion years ago, and it has a life expectancy of another 5 billion years.

At the Sun's birth, in the distant past, there was no one around to take notes. And, lacking a reliable time machine, we have no way of witnessing the Sun's death in the distant future. So how do we know about the beginning and the end of the Sun's life? Even the Sun's aging process is too slow for us to notice—all we have at our disposal is just one snapshot.

Or is that really all we have? After all, the Sun is not the only star we can observe. Suppose you're an alien and your task is to find out about the life cycle of human beings. Unfortunately, your UFO takes off again just one day after it touches down on Earth. During that single day, you don't notice any individual growing older. But if you look around carefully, you see many stages of human life: a baby being born in a hospital, children playing in a schoolyard, a young couple in love, midlife adults fighting wrinkles and love handles, elderly people in wheelchairs, a funeral. Together they paint a vivid picture of the life of a human being.

With stars, it's the same. We don't notice the slow evolution of any particular star. But we can look around the Milky Way galaxy and see stars at various stages of their lives. That's how astronomers have pieced together the story of stellar evolution.

———

Here's the recipe for a star: Take a large amount of gas. Put it in a small enough volume. Then wait. It's as easy as that—nature takes care of all the rest.

The space between the stars isn't really empty. It's filled with gas. In many places, the gas is hot and extremely rarefied—less than one atom per cubic centimeter. Most physicists would call that a perfect vacuum. But elsewhere, cool interstellar gas clouds can be up to a million atoms or molecules per cubic centimeter denser. That's enough for those particles to start feeling some gravitational attraction.

So if a large enough amount of gas sits in a small enough volume of space, gravity will automatically take over. The cloud collapses in on itself for the simple reason that gravity draws the constituent particles as close to each other as possible.

Have you ever tried to put two handfuls of snowflakes as close to each other as possible? You ended up with a snowball, of course. There's no way to pack matter more efficiently than in a sphere. That's the very reason why stars—including our Sun—are spherical. (The same is true for planets, by the way, but not for bricks, mountains, or asteroids—they don't have enough self-gravity to overcome their material strength, which is governed by electromagnetic forces.)

It's easy to see how gravity can pull a tenuous cloud of interstellar gas into a compact sphere. But it's less evident why this gravitational collapse would ever stop. The reason is the pressure of gas in the very core of the newborn star, which exerts an outward force that opposes the inward pull of gravity. The higher the pressure, the harder it becomes to compress the gas any further.

The ignition of nuclear fusion reactions heats up the gas in the star's core and further increases the pressure. The pressure in the core of the Sun, for instance, is some 250 *billion* Earth atmospheres. That's enough to resist the weight of the overlying gas layers—enough to resist gravity. The end result is a star in what astrophysicists call hydrostatic equilibrium. Suppose we could force the star to contract even further. In that case, the density in its core would increase. Nuclear fusion reactions would speed up, giving rise to higher temperatures and pressures. As a result, the star would expand again to its original equilibrium size.

This also means that stars can have—and do have—a wide variety of sizes. The initial diameter of a star depends on the mass of the contracting gas cloud. A larger mass means a higher core density. A higher density means more furious nuclear reactions. More nuclear power means higher temperatures and higher pressures. In the end, hydrostatic equilibrium is reached at a size much larger than the Sun's—nature has cooked up a massive, hot, and luminous giant star.

In contrast, if the original gas cloud is small, the core density remains low. Nuclear fusion occurs at a slow pace, if it commences at all. The interior of the star remains relatively cool; the pressure is not exceedingly high. Hydrostatic equilibrium sets in only when the star has contracted to some 10 percent of the Sun's size—about as large as the planet Jupiter. Presto: a low-mass, cool, and relatively dim dwarf star.

If you think dwarf stars don't matter, think again. First of all, they're much more numerous than their bigger, brighter cousins. Remember, this is nature we're talking about, where small things always outnumber large things. There are more mice than elephants, more pebbles than boulders, more asteroids than planets—it's always like that. But apart from being more numerous, dwarf stars also live much longer than giant stars do.

Wait a moment—they live *longer*? How can that be? If they're so small, don't they have less nuclear fuel available to them than giant stars do? Yes, that's true—their fuel tanks are smaller, so to speak. But dwarf stars are also extremely thrifty. Nuclear fusion occurs at a slow rate and can go on for tens of billions of years despite the relatively small amount of available hydrogen.

If dwarf stars are the slow, economical compact cars of the universe, giant stars are inefficient cosmic gas guzzlers. They may carry much more fuel, but they're big spenders. Before long, they use up their supply of hydrogen—the most massive stars in the universe may live for only a million years or so.

Our Sun is somewhat in between. Not too massive, not too light-weight. As I said before, it's about halfway through its expected lifetime of 10 billion years. But just like any other star, it won't live forever. And because astronomers have seen other Sun-like stars at later stages of their lives, they know when and how the Sun will die.

Over the next few billion years, hydrogen in the Sun's core will start to become depleted as more and more of the stuff is converted into helium. But farther out, in a thick shell around this new helium-rich core, hydrogen fusion is still continuing. As a result, the outer layers gradually expand into space. Our Sun slowly turns into a giant star. That's bad news for life on Earth: less than a billion years from now, the Sun's energy output will already be so high that our planet's oceans will start to evaporate.

Meanwhile, the helium core is growing larger and ever more massive. Helium nuclei are packed closer and closer together. Eventually, some 5 billion years from now, the density is high enough for another round of nuclear reactions to ignite. I'll spare you the quantum-mechanical details, but helium fuses into even heavier elements—first carbon, then oxygen.

Helium fusion produces much more energy than hydrogen fusion does. Because of all that additional energy, the Sun will expand into a bloated red supergiant, well over 100 million kilometers across. Poor Mercury and Venus—those two innermost planets will be devoured, their rocks and metals eventually turned into superheated vapor that gets mixed into the Sun's outer layers. This is planet destruction at the grandest scale.

What about Earth? Well, with a bit of luck, our planet may escape this hellish ordeal. That's because the Sun develops what I would call a stellar fever—a sure sign that the end is near. It starts to pulsate, expanding and contracting again every twenty-four hours or so. As a side effect, the outer hydrogen layers are slowly blown into space. The resulting loss of mass will weaken the Sun's gravitational grip on the planets, causing their orbits to expand. The effect is too small to save Mercury and Venus, but Earth may just survive, its rocky mantle turned into a planet-wide ocean of glowing lava. (Survival is a relative concept.)

Within 10,000 or 20,000 years, most of the Sun's mantle is blown away into a colorful expanding bubble. So far, astronomers have cataloged thousands of similar short-lived bubbles in the Milky Way; there must be many more. For historical reasons, they're called planetary nebulae. William Herschel, who first described them in the late eighteenth century, thought they looked like the circular disks of planets, and the name stuck.

Meanwhile, the explosive fusion of helium comes to a halt. In the blink of a cosmic eye, most of the Sun's helium has been converted into carbon and oxygen. With no energy production left to counteract the pull of gravity, the core starts to contract and continues until what remains is a weird object known as a white dwarf. It packs something like half the original mass of the Sun into a sphere not much larger than Earth. Its density is around 1 kilogram per cubic millimeter.

A white dwarf starts out extremely hot. Its surface temperature may be as high as 100,000 degrees Celsius. But since it has only a small surface area, it doesn't emit a lot of light. Even the nearest known white dwarf, less than 10 light-years from Earth, cannot be seen with the naked eye. Over time, a white dwarf slowly cools down, radiating its residual heat into the cold vacuum of space.

What remains is a dark and inert lump of degenerate matter—a stellar cinder.

RIP, Sun.

So where's the neutron star? Maybe I should have told you up front: the Sun is not massive enough to evolve into a neutron star. White dwarfs are pretty strange as things go, but neutron stars are weirder still. To produce them you need to start out with a star that's at least nine times the mass of the Sun.

As mentioned earlier, massive stars live fast and die young. Their life expectancy is measured in millions of years, not billions. It's as if you speed up the evolution of a star like our own Sun by pressing the fast-forward button on a DVD player. Hydrogen fusion, expanding outer layers, helium ignition, formation of a carbon-and-oxygen core, loss of outer hydrogen mantle—it all happens much more rapidly.

But after that things go very differently. The reason is simple. In a star that is so much more massive than the Sun, the outer layers press heavily on its core. As a result, the density and temperature of the carbon-oxygen core become much higher than in the case of the Sun: more than 3 kilograms per cubic millimeter and some 500 million degrees Celsius. That's enough to start yet another round of fusion reactions, though this time the nuclear engine in the core of the star runs on carbon instead of hydrogen.

Without going into all the details, in about a thousand years (depending on the mass of the star) the carbon is converted into neon, magnesium, sodium, and more oxygen. There's quite a bit of cosmic alchemy going on there! As soon as the carbon is used up, the core of the star begins to contract again. Its density and temperature soar to even higher values—high enough to convert neon into magnesium.

This is when things really start to speed up. After just a few years, most of the neon is also gone. The star's core now consists of oxygen and magnesium. It contracts until oxygen fusion takes over, in which oxygen is converted into silicon and small amounts of sulfur and phosphorus. This process lasts for only a year or so. The stellar core runs out of oxygen, contracts again, and heats up to some 3 billion degrees Celsius. Then, within less than a day, silicon nuclei fuse into a wide range of heavier elements, including argon, calcium, titanium, chromium, and even large amounts of iron and nickel. This is no longer the quiet and steady nuclear fusion process we encountered in the core of our Sun. (Remember, it takes billions of years to slowly convert most of the Sun's hydrogen into helium.) Instead, it's an explosive thermonuclear bomb of astronomical proportions—a cosmic weapon of mass destruction.

If we could slice the stellar time bomb in half, it would resemble an onion. The very core contains iron and nickel—not in the form of solid metals, of course, because it's all gas, albeit at an incredibly high density and temperature. Around the iron-nickel core is a shell of silicon and sulfur. Farther out is another layer containing oxygen, neon, and magnesium. Even higher up are layers of oxygen, carbon, helium, and hydrogen, although by now most of the hydrogen will have escaped into space. Lower-temperature fusion reactions are still active at the boundaries between layers. The stellar onion is brimming with nuclear power. But time is running out.

The catastrophe commences in the core. With all of the silicon gone, the star's nuclear engine is running on empty. The thing is, iron and nickel nuclei do not spontaneously fuse into even heavier elements. Nuclear fusion favors the production of atomic nuclei with higher binding energy (or higher stability), but iron and nickel already have the highest possible binding energy. Simply put, nature sees no reason to convert them into heavier elements.

Gravity immediately seizes the opportunity. For millions of years, gravity has been trying to compress the star to an ever-smaller size, bringing its constituent particles as close together as possible. But over and over again this force has been opposed by the outward pressure of the star's energy production. Now, finally, gravity's patience pays off. The star's nuclear engine shuts down, and no new energy is produced in the core.

Within a second or less, the core of the star collapses in on itself. A few solar masses' worth of incredibly hot gas is compressed into a sphere no larger than some 25 kilometers across—about the size of London or Paris. This superdense ball of nuclear matter, packing about 100,000 tons of mass in every cubic millimeter, is known as a neutron star. So a neutron star is the collapsed core of a massive star that has run out of nuclear fuel.

Why are they called neutron stars? Well, as you might have guessed, it's because they consist of neutrons. I haven't mentioned neutrons before—no need—but now that the subject has come up, let's take a short detour into the world of subatomic particles.

Normal atoms consist of an atomic nucleus surrounded by a cloud of electrons. Electrons are very lightweight particles, so almost all of an atom's mass is concentrated in its nucleus. But an atomic nucleus is not one single particle. It's a collection of protons and neutrons—subatomic particles that have almost the same mass.

The number of protons in an atomic nucleus determines what kind of element we're dealing with. A hydrogen nucleus, for instance, consists of just one proton (and no neutrons at all). Helium has two protons and two neutrons. A carbon nucleus is larger and more massive: it has six protons and six neutrons. Iron has twenty-six of both, so an individual iron nucleus is fifty-two times more massive than a hydrogen nucleus is. Now you understand what astronomers mean by "heavy elements." (For even heavier elements, the number of neutrons in the atomic nucleus is usually somewhat higher than the number of protons. For instance, zinc has thirty protons and thirty-five neutrons.)

Under normal circumstances, the number of electrons surrounding an atomic nucleus is the same as the number of protons in the nucleus: one for hydrogen, two for helium, six for carbon, twenty-six for iron, thirty for zinc, and so on. Since protons have a positive electric charge and electrons have a negative electric charge, this ensures that normal atoms have no net electric charge. (What about neutrons? Well, they're called neutrons for a reason: they are electrically neutral.)

In the core of a star, however, there are no neutral atoms. Conditions are so extreme that electrons are no longer bound to atomic nuclei. Instead, the gas in a star is what's called a plasma: a mixture of charged particles. Positively charged nuclei and negatively charged electrons all go their own way, like parents and children that have lost track of each other in a crowd.

The free-roaming electrons play an important role in the process of nuclear fusion. By interacting with an electron, a proton can turn itself into a neutron. The negative charge of the electron and the positive charge of the proton cancel each other out; what's left is the uncharged neutron. That's how four hydrogen nuclei (four protons) can fuse into one helium nucleus, which consists of two protons and two neutrons. As I've mentioned before, the process also produces positrons (the antiparticles of electrons; those aren't important right now) and neutrinos (spooky elementary particles that are of more importance to our story).

I realize this is a lot of information to digest. What's important to take away is that the collapsing core of a dying giant star contains a plasma consisting of positively charged nuclei of iron and nickel and negatively charged electrons. What's more, the number of electrons is the same as the number of protons in the atomic nuclei.

So what happens when gravity delivers its final blow? The plasma is compressed to unimaginable densities. Individual particles—nuclei and electrons—are forced on each other. In fact, it's fair to say that electrons are violently pushed into the nuclei, which consist of almost equal numbers of protons and neutrons. The electrons can't help but interact with the protons, turning them into even more neutrons. In less than a second, all the protons are gone. What's left is a huge, solid ball of uncharged neutrons packed shoulder to shoulder: a neutron star.

So far we've only talked about the core of the star. What about its onion-like outer layers? Won't they end up in the neutron star, too? No, they won't. To the contrary, the star's outer layers—in fact, most of the star's total mass—blast into space in one of the most dramatic events the universe has in store for us: a supernova explosion.

Why does this happen? As we've seen, at first the whole star starts to collapse. After all, there's almost no energy production going on in the core to counteract the inward pull of gravity. But the free-falling gas—maybe as much as five or six times the mass of the Sun—crashes onto the surface of the newly formed neutron star. Since the neutron star can't be compressed any further, the gas comes to a halt. Its energy of motion is converted into heat, producing a roiling fireball that races outward again, pushing everything in its way ahead of it like a giant bulldozer.

The neutrinos I mentioned earlier also play a role here. Remember, neutrinos are produced when protons interact with electrons to change into neutrons. The formation of the neutron star creates a tsunami of neutrinos—one for every newly produced neutron. Although neutrinos hardly interact with normal matter, they provide an additional outward push. The overall result: while the stellar core collapses into a compact ball of neutrons, most of the star is blown into smithereens, violently exploding into space in a rapidly expanding shell.

A supernova is serious business. For weeks on end, the catastrophic explosion can produce more light than all the stars in a galaxy put together. Personally, I'll never forget supernova 1987A, which exploded in late February of that year in the southern sky. Three months later I visited the European Southern Observatory in Chile for the first time in my life. The fading light of the stellar explosion was still easily visible to the naked eye—quite impressive, given that it came from 167,000 light-years away.

You don't want a nearby star to go supernova—its energetic radiation would blow away Earth's atmosphere and sterilize our planet. Fortunately, supernovae are relatively rare. The last one observed in our own galaxy took place in 1604, and it occurred at a safe distance of some 20,000 light-years.

So there you have it. Neutron stars, which will be so important to our story of gravitational waves, are the ultraweird mortal remains of stellar giants. (As for weirdness, we've only scratched the surface so far; there's much more to follow later on in the book.) And a neutron star's formation is accompanied by one of the most titanic explosive events in the universe: a supernova. In Chapter 6 we'll see how neutron star observations in the 1970s confirmed the existence of Einstein waves long before the minute spacetime undulations were directly detected.

———

Oops, I completely forgot Carl Sagan's apple pie. Sorry about that— I got carried away by the exciting story of stellar evolution. But of course, Sagan's *Cosmos* quote—"If you want to make an apple pie from scratch, you must first invent the universe"—*is* about cosmic evolution. Without the formation of galaxies, without the birth of stars, without planetary nebulae and supernova explosions, that apple pie never could have been baked.

The Crab Nebula, in the constellation Taurus, is the expanding remnant of a supernova explosion that was observed by Chinese and Korean astronomers in AD 1054. In the center of the nebula is a rapidly spinning neutron star—the contracted remains of the stellar core.

As we'll see in Chapter 9, the universe began as a primordial soup of elementary particles. A few hundred thousand years later, these had collected into simple atoms of hydrogen and helium. If there had never been stellar evolution, if the nuclear ovens of the cosmos had never been fired up, that would still be all there is—hydrogen and helium. Not much to work with.

Apple pies—just like chairs, cats, and car keys—contain large amounts of heavier elements. Carbon, oxygen, and nitrogen. Sodium, calcium, and phosphorus. Magnesium, aluminum, and iron. And these have all been cooked up in stellar interiors over the past 13.8 billion years of cosmic evolution. Together, they constitute a mere 1 percent or so of the total atomic mass of the universe, but that small amount makes all the difference.

Blown into the void by stellar winds, in planetary nebulae, and through stellar explosions, these elements slowly worked their way into interstellar space. Small amounts of even heavier atoms, such as copper, zinc, gold, and uranium, were crafted in the helter-skelter aftermath of supernovae or during catastrophic collisions of neutron stars. Gas clouds became enriched with complex molecules and dust particles. New generations of stars found themselves accompanied by planets, some warm enough to harbor liquid water. On at least one of those rocky worlds, carbon-bearing molecules rained down and eventually became organized into the first living organisms. A few billion years later, the planet boasted wheat, sugarcane, and apple trees—the necessary ingredients for apple pie.

And there were people.

For what's true of apple pies is also very true of you and me. In my humble opinion, this is the most wonderful story that science has to tell: that the carbon in your muscles, the calcium in your bones, the iron in your blood, and the phosphorus in your DNA have all been synthesized by nuclear fusion reactions in distant suns. As Canadian folk singer Joni Mitchell wrote in her 1969 ballad "Woodstock," "We are stardust—billion-year-old carbon."

The lives of stars are intimately connected to the lives of ourselves. We are one with the cosmos.

(((6)))

Clockwork Precision

Pulsar is the name of an American brand of watch. The company is part of the Seiko Watch Corporation. In 1972, Pulsar built the first LED watch. Electronic. Digital. *Very* cool (mind you, this was forty-five years ago).

Pulsar is also the name of a hatchback first produced in 1978 by the Japanese car company Nissan. A popular sports motorbike model made by Bajaj Auto Ltd. in India carries that nameplate. A high-tech lighting company in the United Kingdom is called Pulsar, as is a manufacturer of night-vision gear in Lithuania.

But before 1967, the word *pulsar* did not exist. It first appeared in print in the UK newspaper the *Daily Telegraph* in the spring of 1968. That story wasn't about watches, cars, motorcycles, lighting, or night vision. It was about a surprising astronomical discovery. Ten years later, that discovery led to the first indirect detection of gravitational waves.

In Chapter 5, you read about neutron stars. They are the leftovers of supernova explosions—the mortal remains of massive stars that have blown themselves to pieces. Very small and incredibly dense, neutron stars are among the most extreme denizens of the universe. Their existence had first been predicted in 1934 by Walter Baade and Fritz Zwicky, two European astronomers who had emigrated to the United States, just like Einstein.

Supernova explosions must have been occurring in our galaxy for billions of years. So in the 1960s, astronomers were well aware that there should be tens of millions of neutron stars in the Milky Way. But they had never observed one. That's not too surprising, if you think about it. While the surface of a newborn neutron star is extremely hot, it has an area of just a few hundred square kilometers. As a result, the total amount of high-energy radiation is rather small. Even a nearby neutron star would be hard to spot.

So the discovery made by twenty-four-year-old graduate student Jocelyn Bell came as a surprise. Born in Northern Ireland, Bell worked at Cambridge University, England, under the supervision of radio astronomer Antony Hewish. In the 1960s, radio astronomy, which studies long-wavelength radiation from all corners of the universe, was still a relatively young field, and new discoveries were being made all the time.

The radio telescope that Bell had helped build was basically an array of wooden poles connected with wires—a bit like an old-fashioned TV aerial, only much larger. The low-cost contraption picked up radio waves from the sky and each day produced about thirty meters of chart recordings that more or less resembled the output of a seismograph.

This all occurred in the summer of 1967, which is generally known as the Summer of Love. Hippies smoked pot in San Francisco's Haight-Ashbury neighborhood; the Beatles were recording their album *Magical Mystery Tour.* Meanwhile, Jocelyn Bell spent most of her time visually inspecting the radio telescope's paper records to see if she could find anything unexpected in the wiggly patterns.

Which she did, in August.

Long story short, Bell found a mysterious, pulsating radio signal originating in the small constellation Vulpecula, the Fox. It produced a brief beep every 1.3 seconds, like a cosmic metronome.

You may have heard the story, and it's indeed true: for a number of weeks, Bell, Hewish, and their colleagues considered the possibility that they might have found aliens. What natural phenomenon would be able to produce such a rapid, extremely regular radio signal? It certainly seemed like it was artificially created, yet it was clearly

Twenty-four-year-old Jocelyn Bell poses in front of the Cambridge radio telescope with which she discovered the first pulsar in 1967.

extraterrestrial. They even code-named the signal LGM-1, for "little green men."

Surprisingly, Bell was annoyed by this. You would think a young astronomer would be elated by the possibility that she'd found evidence of extraterrestrials, but no. "Here was I trying to get a PhD out of a new technique, and some silly lot of little green men had to choose my aerial and my frequency to communicate with us," she told the audience during an after-dinner speech at a December 1976 conference in Boston.

The LGM phase didn't last long. Within a couple of months, Bell had found three additional similar pulsating radio sources in very different parts of the sky. There was no chance that four unrelated alien civilizations would use the same type of communication. There had to be a natural cause. A paper announcing the discovery was published on February 24, 1968, in *Nature.* The lead already hinted at a possible explanation: "The radiation seems to come from local objects within the galaxy," it read, "and may be associated with oscillations of white dwarf or neutron stars."

Not much later, when interviewed by the *Daily Telegraph,* Hewish first used the word *pulsar*—short for "pulsating star."

But why would a neutron star produce regular pulses of radio waves?

Here's the idea (and it's not oscillations, as the *Nature* paper suggested). Neutron stars are not only unbelievably dense, they're also rapidly rotating. That's because of the conservation of angular momentum, but let's just call it the ice-skater effect. Have you ever seen Russian figure skater Evgeni Plushenko? He won four Olympic medals, as well as the 2001, 2003, and 2004 World Figure Skating Championships. If you watch him draw in his arms during a spin, you'll see his rate of rotation increase. It's a law of nature: spinning things that shrink spin faster. (If you're not good at figure skating, you can still experience the effect yourself. Sit down on an office chair. Spread out both your arms and legs. Ask someone to turn you around as fast as possible. Then draw in your limbs. There you go.)

The slowly rotating core of a massive star that collapses into a ball of neutrons less than 25 kilometers across is the astrophysical counterpart of Evgeni Plushenko: the spin rate increases dramatically. Newborn neutron stars can rotate many times per second.

The collapse of the stellar core has another effect, too: the strength of its magnetic field will increase dramatically. Neutron stars have magnetic fields that can be at least a hundred million times stronger than the Earth's. So our small, dense ball of neutrons is a highly magnetized, rapidly spinning cosmic top. Certainly not your average star.

Now it gets interesting. Rotating magnets produce currents—if your bicycle has an old-fashioned dynamo, you know that. Currents are flows of charged particles, and accelerated charges produce light and other forms of electromagnetic waves, as Maxwell has taught us. In other words, magnetized neutron stars emit electromagnetic radiation, primarily along their magnetic axis. From both the north and south magnetic poles of the neutron star, powerful beams of radio waves, light, and even X-rays are funneled into space. (Note what I just said: from the north and south *magnetic* poles. In most cases, those will not coincide with the north and south *rotational* poles. This is the case with our Earth, too.) So as the neutron star rotates, the narrow beams of radiation sweep through space, much like the beams of a lighthouse. If your radio telescope happens to lie in the path of one of the beams, you'll detect brief radio pulses once per revolution. The neutron star has exposed itself as a pulsar. (And yes, some pulsars have also been observed to emit pulses of light and / or X-rays.)

So thanks to this lighthouse effect, pulsars *can* be observed after all, provided you're in the right location. Jocelyn Bell's discovery during the Summer of Love marked the very first observation of a neutron star since Baade and Zwicky had predicted their existence thirty-three years before. And the frequency of the radio pulses (one pulse every 1.3373 seconds) immediately told astronomers the rotation period of the neutron star. That's *fast,* by the way. Imagine something the size of London or Paris rotating three times every four seconds.

Great stuff. That's what radio astronomer Joe Taylor thought when he first heard about pulsars. He was twenty-six when he read the *Nature* paper. At Harvard University in Cambridge, Massachusetts, Taylor had just completed his PhD thesis on the occultation of radio sources by the Moon. But pulsars seemed much more exciting to pursue—not by visually inspecting reams of chart paper, the way Bell had done, but through a systematic, automated search. Off he went to the National Radio Astronomy Observatory in Green Bank, West Virginia. Within a year, Taylor and his colleagues had found six more pulsars. The hunt was on.

I'm sure Albert Einstein would have loved pulsars. Part of Einstein's theory of general relativity is about the influence of strong gravitational fields on the flow of time. The gravitational field at the surface of a neutron star is something like a few hundred billion *g*'s—a few hundred billion times more than what's experienced by a falling apple on Earth. Moreover, pulsars are very accurate clocks (enormously more precise than their watch namesakes). You couldn't wish for a better laboratory to study the effects of general relativity. No wonder astronomers wanted to catch as many as they could.

However, finding pulsars is easier said than done. Most radio telescopes have an extremely small field of view. You don't know in advance where to look. You also don't know what pulse period to search for in the data. Moreover, at lower radio frequencies, the pulse signal arrives later than at higher frequencies. The reason for this is that radio waves are a bit slowed down by the smattering of electrons in the almost-empty space between the stars, and this effect is stronger at lower frequencies. So if you're observing at a certain range of frequencies, which is usually the case, the pulses are smeared out. It's what radio astronomers call "dispersion." Pulses may stand out against the ever-present background noise only if you correct for this effect. But the amount of dispersion is dependent on the pulsar's distance: for more distant pulsars, there are more intervening electrons. And since you don't know the distance of an as-yet-undiscovered pulsar, you don't know which dispersion measure to correct for.

Still, by 1974, discovering new pulsars had become nearly routine—at least for Russell Hulse, a graduate student at the University of Massachusetts at Amherst, where Taylor had moved in 1969. Hulse's job: search the Milky Way and find as many new pulsars as he could. His instrument: the 305-meter radio telescope at the Arecibo Observatory in Puerto Rico, which later became famous through movies such as *GoldenEye* (1995) and *Contact* (1997). His weapon: brute force.

Hulse spent most of 1974 at Arecibo, coping with heat, humidity, and mosquitoes—and with the idiosyncrasies of his 32 KB minicomputer, a novelty in those days. For a few hours each day, when the Milky Way rose high above the giant Arecibo dish, he collected new radio data. Then he fed all the data into his computer. Dedicated software searched for the presence of rapid pulses by trying no less than half a million possible combinations of different pulse periods and dispersion measures. Every now and then, the search would pay off. On average, Hulse found about one new pulsar every ten days. I figure colleagues must have called him "Russell Pulse" back then.

The 305-meter radio telescope near Arecibo, Puerto Rico, is built in a bowl-shaped valley. Using this giant instrument, Russell Hulse discovered the first binary pulsar, PSR B1913+16, in 1974.

The big surprise came in the summer of 1974—around the time of the Watergate scandal. Hulse had found a particularly fast pulsar at a distance of some 20,000 light-years. It spun around once every 59 milliseconds, producing seventeen extremely brief radio pulses per second. At the time, it was the second-fastest pulsar known, which made it quite interesting already. But a couple of weeks later, when he observed the pulsar again, he found something strange: the pulse period had changed. Not much—less than 1 / 10,000 of a second—but still. Later, it changed again, this time in the other direction. Hulse found that extraordinary. Weren't pulsars supposed to be the most accurate clocks in nature? How could a massive, supercompact spinning ball of neutrons suddenly speed up or slow down?

Eventually it dawned on Hulse that the pulsar had to be part of a stellar binary. If it was orbiting another star, one he couldn't see, it would alternately move toward us and away from us. Approach, recede, approach, recede. When the pulsar moves in our direction, its radio pulses arrive at Earth slightly closer in time—a higher pulse frequency. When it moves away from us, its pulses arrive slightly farther apart in time—a lower frequency. Russell Hulse had discovered the first known pulsar in a binary system.

The frequency change that Hulse observed is known as the Doppler effect. It's what happens to the sound of an ambulance's siren as the vehicle speeds by. When the ambulance is racing toward you, you experience the siren's sound waves as compressed, and you hear a higher-pitched tone. As the ambulance speeds away, you experience the waves as being stretched out, resulting in a lower sound.

The Doppler effect is named after nineteenth-century Austrian astronomer Christian Doppler. In 1842, he suggested the phenomenon might explain the striking color difference in some binary stars. The light from an approaching star observed at a higher frequency would be a bluer color, while a receding star would appear redder, corresponding to a lower frequency. Doppler was wrong about this: the colors of stars are determined by their surface temperatures, not by their movement through space. Stars would have to be traveling at pretty close to the speed of light to visibly change hue. Yes, orbiting binary stars *do* exhibit a small frequency (or wavelength) change, but it's not visible to the eye, and you need very sensitive measuring devices to detect it.

Three years later, in 1845, Dutch meteorologist Christophorus Buys Ballot was the first to demonstrate the Doppler effect for sound, using not an ambulance, but a train. The rail line between the Dutch cities of Amsterdam and Utrecht had just been established, and Buys Ballot arranged for a locomotive to travel back and forth along the track around the train station at Maarssen, a small village just seven kilometers northwest of Utrecht, as horn players on the train and on the platform played the same tone. The Doppler effect was very evident—you didn't need perfect pitch to notice the frequency difference. (I like this story a lot because I grew up in Maarssen, just a few hundred meters from the railway station.)

What's so special about a pulsar in a binary system? For one thing, it helps you determine the mass of the neutron star, something that's essential to understanding the true nature of these unusual objects. Moreover, knowing the mass and the precise orbit of the neutron star within the system would help to test some of the predictions of Einstein's theory of general relativity. All this information can be gained from carefully studying the arrival times of the radio pulses.

Remember the conservation of angular momentum, aka the ice-skater effect? It explains why Evgeni Plushenko spins faster when he draws in his arms. It also guarantees that massive, rapidly rotating bodies keep on spinning unless acted upon by some force.

In Plushenko's case, the main braking force is the friction of his skates on the ice. Without friction (and air resistance), his spin would never end. Neutron stars don't have skates, and in the vacuum of space, there's no air resistance. Moreover, neutron stars are far more massive than the average ice-skater, which makes it much harder to slow them down in the first place. The result of all this is that a rapidly rotating neutron star basically keeps on spinning forever at exactly the same rate. (For the nitpickers among you, there's some magnetic braking, but it is extremely slow—you wouldn't notice it in the course of a human lifetime.)

But if the spin rate of a neutron star never changes, all peculiarities in the pulse arrival times must be attributable to some other physical effect. It's just a matter of measuring, analyzing, untangling, deducing, and checking.

The Doppler effect found by Hulse is the easiest part. Hulse saw how the pulse frequency increased and then decreased over a period of seven hours and forty-five minutes. If this is due to the pulsar's orbital motion, it follows that the orbital period is also seven hours and forty-five minutes. (To be very precise, it's seven hours, forty-five minutes, and seven seconds.) Bingo—there's our first orbital parameter.

If the orbit were a neat circle, the observed pulse frequency would change gradually and symmetrically. But it doesn't. On average, the frequency is 16.94 pulses per second (corresponding to a spin rate of 59.03 milliseconds). For about five hours each orbit, the observed frequency is lower, meaning the pulsar is receding from us. For the remaining two and three-quarter hours, the observed frequency is higher, which means the pulsar is approaching. Not symmetric at all. This immediately tells you that the orbit can't be a circle. It must be highly eccentric. (For the record, the orbital eccentricity is 0.617.) That's our second piece of information.

Taylor and Hulse also found that the pulsar's orbit can't be much larger than about 1 million kilometers across. When the pulsar is on the far side of its orbit (as seen from Earth), the pulses arrive some three seconds late compared to when it's on the near side. Radio waves travel at the speed of light (300,000 kilometers per second), so three seconds correspond to almost 1 million kilometers. (This is of course the *projected* size, measured along the line of sight. If the orbit is tilted, the true size must be larger.)

What's next? Well, the timing measurements revealed that the eccentric orbit itself is precessing—and pretty fast, for that matter. Remember the perihelion precession of Mercury? Urbain Le Verrier found that it was larger than he would have expected based on Newton's theory of universal gravity. Einstein was able to explain the observed excess of 43 arc-seconds per century as due to the curvature of spacetime. But this relativistic effect is much larger for the pulsar's orbit: more than 4 degrees per year. That means the pulsar's orbit precesses as much in one day as Mercury's orbit does in about a year. And *that* can mean only one thing: very strong spacetime curvature caused by a very strong gravitational field.

And there's more. Pulsars are nature's perfect clocks. A pulsar orbiting a binary companion is like an atomic clock orbiting the Earth. It's an astrophysical version of the Hafele-Keating experiment described in Chapter 3, albeit without the flight attendant. Sure enough, the kinematic time dilation betrayed itself in the observed pulse arrival times. Of course, the effect is much larger than Hafele and Keating had measured because of the pulsar's high orbital velocity, which varies between 110 and 450 kilometers per second. That's about a thousand times as fast as your average airliner and about one-thousandth of the speed of light.

Doppler effect, eccentricity, orbital precession, time dilation—each effect yielded a new piece of knowledge. Put all those together, and you can deduce other things you did not yet know. The tilt of the orbit, for instance, which is some 45 degrees. Or the true spatial distance between the two orbiting stars, which varies between 746,600 and 3,153,600 kilometers. And, most important, the masses of the two objects: the pulsar itself is 44.1 percent more massive than the Sun, a very typical value for a neutron star, but the companion object is almost as hefty, 38.7 percent more massive than the Sun. Could the companion be a normal star? No way, because if it was, it would also be much larger than the Sun—too large, in fact, to fit within the pulsar's orbit.

Small, massive, and invisible, even to the largest telescopes—what kind of object could it possibly be? You probably guessed it: another neutron star, and one that is not oriented in the right direction to be observed as a pulsar, at least not from Earth. Alien astronomers on some remote planet may find themselves observing the lighthouse-like beams of this second pulsar (if it emits beams of radiation at all). To them, *our* pulsar would be invisible.

Another thing to realize is that most alien astronomers would not be able to observe the system at all, as they would be outside the sweeping beams of radiation of both pulsars. We're just lucky. Then again, there must be many binary neutron stars in the Milky Way that *we* cannot observe. They may pulse like crazy, just not in our direction.

———————

All in all, it's an impressive piece of detective work. All you've observed is the *beep-beep-beep* from one pulsar. But for a smart cosmic Sherlock Holmes, that's enough. Just carefully analyze the minute deviations from perfect regularity. That tells you everything you need to know about this fascinating binary system. Plus it lets you check on the predictions of Einstein's theory of general relativity. (As you might expect, general relativity passed the tests with flying colors.)

After Hulse left the University of Massachusetts at Amherst in 1975, Taylor continued the investigation with Joel Weisberg, a graduate student at the University of Iowa who was later recruited by Taylor as a postdoc. Together, they made the discovery of a lifetime.

Taylor and Weisberg realized that if Einstein's predictions are borne out, the binary pulsar should lose energy. Here are two massive, compact objects orbiting each other at breakneck speed. General relativity says these accelerating masses should produce ripples in the fabric of spacetime—gravitational waves. The waves would carry energy away. As a result, you'd expect the two orbiting neutron stars to lose orbital energy. Slowly but surely, they would spiral in toward each other. The orbit must shrink; the period has to decrease.

The masses and the orbit of the binary neutron star are known to high precision. Enter those values into Einstein's equations, and out pops the prediction for the orbital decay. In one year, the average distance between the two neutron stars should shrink by 3.5 meters. That's hard to measure from a distance of 20,000 light-years, as you can imagine. But the corresponding decrease in the orbital period is 76.5 microseconds per year. And that should be evident in the pulse arrival times, at least after a couple of years.

And it was. In 1978, Taylor, Weisberg, and their colleagues found that their results perfectly matched the predictions of general relativity. Einstein *was* right. They announced their finding at the ninth Texas

Symposium in Munich, Germany, in December of that year; two months later, the discovery was published in *Nature*. The message was clear: the shrinking orbit of the binary pulsar was proof—indirect, but very convincing—of the existence of Einstein waves.

That's what the Nobel Committee thought, too. In November 1993, the Nobel Prize in Physics was awarded "for the discovery of a new type of pulsar, a discovery that has opened up new possibilities for the study of gravitation." The prestigious prize was shared between Joe Taylor (who had moved to Princeton University in 1981) and Russell Hulse.

What about Joel Weisberg? Why didn't he share in the Nobel? True, he hadn't discovered the binary pulsar. But then again, by the time Taylor and Weisberg discovered the Einstein wave effect, Hulse was working in plasma physics—a completely different field. *He* didn't discover the decay of the pulsar's orbit. Moreover, Nobel Prizes can be awarded to a maximum of three individuals. Weisberg could easily have been the third laureate. So why wasn't he included?

For some reason, the Nobel Committee appears to have a complicated relationship with work involving pulsars. In 1974, the year Hulse found the binary pulsar, the committee awarded half of the Nobel Prize in Physics to Antony Hewish "for his decisive role in the discovery of pulsars." That sort of makes sense, if "decisive role" means hiring the graduate student who actually made the discovery. But the pioneering work of Jocelyn Bell wasn't even mentioned.

These days, Weisberg is at Carleton College in Northfield, Minnesota, and even though his work wasn't acknowledged by the Royal Swedish Academy of Sciences, he is happy that the discovery received the appreciation it deserves. And he is still monitoring the Hulse-Taylor pulsar, as it is now generally known. Over the years, the measurements have become ever more precise. Deviations from Einstein's predictions have not yet surfaced. And Weisberg is studying other binary pulsars, too. By now, dozens of them have been found. They provide astrophysicists with free gravitational laboratories in space: rent a radio telescope, hook up a timing device, and you're in business.

One of the most exciting systems is PSR J0737–3039. It was discovered in 2003 by Italian radio astronomer Marta Burgay, using the 64-meter radio telescope at the Parkes Radio Observatory in Australia. The numbers in its name are like a celestial address: they represent the pulsar's location in the sky in the southern constellation of Puppis. (Likewise, Jocelyn Bell's original pulsar is officially known as PSR B1919+21. The Hulse-Taylor pulsar, in the constellation of Aquila, is PSR B1913+16. As you can tell from the numbers, those two pulsars are not too far from each other in the sky.)

What makes J0737 so special? Well, it's the only known genuine double pulsar. Although the first binary pulsar was a double neutron star, only one of the neutron stars is seen to pulse. But with J0737, both neutron stars are observed as pulsars. Moreover, they're in a very tight orbit: high velocities, strong accelerations. This all provides for more precise measurements and additional cross-checks on the results.

And there's one more thing that's unusual about J0737: the orbital plane of the two whirling pulsars is seen almost edge-on. Every 1.2 hours (half the orbital period), one is seen more or less behind the other, and its beam passes very close to the other pulsar when observed from Earth. Because of the strong gravitational field, time slows down (gravitational time dilation), and the signal takes more time to reach our radio telescopes than it would otherwise. This lag, called the Shapiro delay, has been measured to great precision. It's exactly as large as predicted by general relativity.

Irwin Shapiro could never have guessed that the general relativity test named after him would be applied to binary pulsars. Shapiro was an astrophysicist at MIT in 1964; when he first described the effect, pulsars hadn't been discovered yet. Shapiro suggested bouncing radar signals off the planets Mercury and Venus around the time of superior conjunction, which occurs when the planet is located on the far side of the Sun as seen from Earth. The radar signals would have to pass through the gravitational field of the Sun. Precise measurements of the travel time of the radar pulses would reveal the resulting time delay.

The first such radar experiments, carried out by Shapiro and his colleagues in 1967, weren't too precise. But yes, the delay was measured, and yes, the results agreed with Einstein's predictions. More recently, the Shapiro delay has also been measured (to much higher precision) in radio communication signals from NASA's space probe Cassini, which has been orbiting the planet Saturn since 2004. But the latest observations of PSR J0737–3039 yield an even higher accuracy.

Another exciting binary pulsar is PSR J1906+0746, found in 2004 with the Arecibo radio telescope. It rotates once every 144 milliseconds, producing almost seven radio pulses per second—nothing too special (you can eventually get used to anything, even super-dense, city-sized stars spinning as fast as the wheels of a speeding car). But in 2008, the pulses became weaker and weaker. By 2015, they had disappeared altogether. Now *that* is remarkable.

Or is it? In fact, the fading behavior of PSR J1906+0746 is explained by general relativity, too. It's caused by geodetic precession, also known as de Sitter precession. Thanks to the strong curvature of spacetime, the orientation of the pulsar's spin axis is slowly changing over time. (The same effect was detected by Gravity Probe B, as I described in Chapter 3.) The magnetized cosmic top is wobbling. As a result, its lighthouse beams of radio waves no longer sweep over the Earth. The pulsar disappears, at least for us. Luckily, it is predicted to reappear again around the year 2170. Future radio astronomers, mark your calendars. (By the way, both geodetic precession and Shapiro delay have also been discovered in the Hulse-Taylor pulsar, in 1989 and 2016, respectively.)

We've come a long way since "some silly lot of little green men" almost ruined Jocelyn Bell's PhD project. Half a century of astronomical detective work has yielded more than 2,000 pulsars in our Milky Way galaxy, including dozens of pulsars in binary systems. This is wonderful material for astronomers who want to understand the final stages of the evolution of massive stars. It is revealing data for nuclear physicists, who investigate the behavior of matter under extreme densities. And it is great stuff for the heirs of Albert Einstein: there's no better way to reveal the secrets of spacetime than by studying these cosmic gravitational laboratories.

For our story, of course, the orbital decay of binary pulsars is the most important result. The fact that the orbital period of the Hulse-Taylor pulsar decreases by 76 microseconds per year is indirect proof of the existence of gravitational waves. To refresh your memory, accelerating masses produce ripples in the fabric of spacetime. These Einstein waves carry away energy. The energy loss of the binary system causes the orbit to shrink. It's as simple as that.

In case you're wondering about the amount of lost energy, it's huge. Every single second, the Hulse-Taylor pulsar loses 7.35×10^{24} joules. That's about a thousand times the energy that was released 66 million years ago when a 10-kilometer asteroid hit the Earth and wiped out the dinosaurs. *Per second.*

If so much energy is pumped into spacetime, the resulting Einstein waves must be huge. At least that's what you'd think. But no, they are tiny. Incredibly tiny. Remember the comparison I made between tapping a bowl of Jell-O and sledgehammering a block of concrete? Well, spacetime is unimaginably stiff. Even the energy of a thousand killer asteroid impacts per second isn't able to set off noticeable ripples.

By the way, the gravitational waves from the Hulse-Taylor pulsar are really low-frequency waves. The orbital period of 7.75 hours implies a frequency of about 72 microhertz. The corresponding wavelength is an incredible 4.2 billion kilometers. So we're talking extremely long-wavelength, low-frequency, small-amplitude ripples. Could we ever hope to measure them? No way, and certainly not from a distance of 20,000 light-years.

But things will improve in the future. The two neutron stars are spiraling in toward each other, slowly but surely. As they get closer and closer together, the orbital period gets smaller and smaller. Binary systems always produce two gravitational waves per orbit, so the frequency of the spacetime ripples will increase over time. And the amplitude of the waves will increase, too, as the neutron stars end up in ever-smaller orbits and experience ever-stronger accelerations. Shorter wavelengths, higher frequencies, larger amplitudes—if we're patient enough, we might be able to directly detect Einstein waves from the Hulse-Taylor pulsar. That's good news.

The bad news is that we'll need a lot of patience. The waves won't be measurable until the two neutron stars are whirling around furiously just a few tens of kilometers apart. Right before they collide and merge—probably transforming into a black hole—the frequency and amplitude will rise dramatically. The coalescence will produce a final, powerful burst of Einstein waves that could be picked up by detectors here on Earth—something already predicted in 1963 by Princeton physicist Freeman Dyson. But in the case of the Hulse-Taylor pulsar, this won't happen until some 300 million years from now.

But wait. Other binary systems show the same behavior: shrinking orbits, decreasing periods, and eventually collisions. For instance, PSR J0737–3039 (the famous double pulsar) will merge in about 85 million years. WD 0931+444 (a system consisting of two orbiting white dwarf stars) has a remaining lifetime of less than 9 million years. Another white dwarf binary, J0651+2844, will coalesce a mere 2.5 million years from now. Who knows, our Milky Way galaxy may contain systems that will collide ten years from now. Or tomorrow. Remember, there must be large numbers of neutron star binaries that we can't observe because their lighthouse beams sweep around in the wrong direction.

Furthermore, we don't need to limit ourselves to our own Milky Way. The final merger of two massive, compact objects such as neutron stars or white dwarfs produces really powerful Einstein waves—so powerful that they could be detected here on Earth even if the collision occurred in a neighboring galaxy. Build a sensitive gravitational wave detector, and you may be able to detect the spacetime ripples from neutron star mergers out to a distance of tens of millions of light-years.

Interestingly, such remote neutron star mergers may already have been observed. Every now and then, Earth-orbiting satellites detect brief explosions of energetic gamma rays from deep space. These gamma-ray bursts (more on them in Chapter 14) come in two varieties. The long ones, with durations of many seconds or even minutes, probably result from exploding supermassive stars. But the short ones, lasting a mere fraction of a second, are most likely produced by neutron star mergers in distant galaxies.

Anyway, the discovery of pulsars and the detection of the orbital decay of compact binary systems enormously inspired and sped up the hunt for gravitational waves. As Joel Weisberg, Joe Taylor, and Lee Fowler wrote in a 1981 article in *Scientific American,* "The binary-pulsar experiment should encourage the investigators who are developing gravitational-wave experiments. It now seems assured that what they are looking for does in fact exist."

Yes, gravitational waves exist.

Yes, neutron stars collide.

Yes, we should try to finally detect those elusive spacetime ripples directly.

If resonant bar antennas are not up to the task, it's time to try a different approach, a new technology that's much more sensitive than Joe Weber's aluminum cylinders: laser interferometry.

$(((\ 7 \ )))$

Laser Quest

I've visited LIGO twice.

The first time was in the spring of 1998. Back then, the Laser Interferometry Gravitational-Wave Observatory was still under construction. It was basically a large empty building plus two half-finished steel pipes 1.2 meters in diameter. Site manager Gerry Stapfer showed me around, but there wasn't much to see yet. "This is where the control room will be"—a large empty room with a number of still-unpacked boxes. "Here are the offices"—smaller empty rooms with plastic-wrapped furniture. "And here's the LVEA," the Laser and Vacuum Equipment Area—wow, a huge empty hall, *really* huge. A forklift truck on the other side looked like a toy vehicle. A small circle on the concrete floor marked the future location of the heart of LIGO: the beam splitter.

My second visit was in late January 2015, some thirteen years after LIGO started its quest for gravitational waves. That was a very different experience, as you can imagine. Laser light bounced back and forth through the two 4-kilometer arms of the interferometer. It took me about ten minutes to drive my car from one end of the L-shaped detector to the other (the sheer size of the facility is evident from the air or from Google Earth). In the control room, young scientists and

Aerial view of the Laser Interferometer Gravitational-Wave Observatory (LIGO) at the Hanford site in Washington State.

engineers had their eyes glued to computer monitors. Hipster beards, ponytails, and nerdy T-shirts abounded. Giant screens on the wall revealed the status of the detector's instruments. The LVEA was now chock-full of delicate instrumentation stowed away in stainless-steel vacuum tanks. A nice set for a James Bond movie.

What was also very different was the view from the roof of the main building. In 1998, I looked out over the woods and wetlands near Livingston, Louisiana. In 2015, I was treated to a panorama of the arid Hanford Site in southeast Washington State. Before you start wondering, I'll remind you that there are two identical LIGOs, some 3,030 kilometers apart. (Why? For the same reason Joe Weber operated two widely separated bar detectors—an attempt to eliminate false positives.) As long as you stay inside, though, you wouldn't notice the difference between the two facilities. When Livingston scientists visit the Hanford observatory, they have no problem at all finding their way around (although I've been told some doors open in a slightly different way).

The Department of Energy's Hanford Site, north of the city of Richland, is not a popular tourist destination. Here, more than seventy years ago, a plutonium reactor produced the nuclear fuel for the atomic bomb that exploded over the Japanese city of Nagasaki in August 1945. Northwest of LIGO, Geiger counters betray the presence of huge amounts of nuclear waste stored in underground bunkers. Route 10, connecting Highway 240 and Glade North Road, is a long, straight strip of tarmac in the desert. Dust and tumbleweed blow across LIGO's short access road.

In Louisiana, the atmosphere is very different. Livingston, east of Baton Rouge, is a small, quiet town. There is a gas station, a hardware store, and a few hundred houses—that's about it. A turn at the Fireworks Warehouse USA takes you onto Highway 63 going north. The road meanders gently through the forest for a while, and then you turn onto a dirt road that leads northwest to the observatory. The central building is surrounded by small ponds and groups of trees. It's all more laid-back, as you would expect in the Pelican State.

This is where scientific history was written very early in the morning of Monday, September 14, 2015, at 04:50:45 Central Daylight Time (Livingston) or 02:50:45 Pacific Daylight Time (Hanford), to be precise. A century after Albert Einstein was completing his theory of general relativity, the twin LIGO observatories made the first direct detection of a passing gravitational wave. For about one-fifth of a second, the sensitive detectors measured ripples in spacetime that were 10,000 times as small as the diameter of a proton—the nucleus of a hydrogen atom. A decades-long search had finally paid off.

We'll get back to GW150914 later in this book, and I'll tell you more about LIGO's checkered history in Chapter 8. But first let's focus on the technology itself. It sounds like magic. One ten-thousandth of an atomic nucleus—how would someone go about measuring such an incredibly tiny effect? And how can you be sure it's really Einstein waves you're detecting and not something more mundane?

Let's start with the basics. What exactly are we trying to measure? Ripples in spacetime. I introduced the concept in Chapter 4—feel free to reread the relevant paragraphs there if you need to refresh your memory—but here's the main message: Draw a large square on the ground. A perpendicular gravitational wave arriving from the point right above your head (the zenith) will slightly deform the square. First it will grow in the north-south direction and shrink in the east-west direction. Next the square is squeezed north-south and stretched out east-west. The square is shivering. How fast? Depends on the wave's frequency. How big are the deformations? Depends on the wave's amplitude.

So all we have to do is precisely monitor the dimensions of a square, preferably in two directions at the same time. No need to measure all four sides of the square, of course. It's okay to just keep an eye on two perpendicular sides that meet each other in one of the four corners. There's your big L shape. Now you know why LIGO looks the way it does.

What if the gravitational wave doesn't come from straight above? Well, the two arms of the L will still grow and shrink, but to a lesser degree, depending on the angle of incidence. But yes, LIGO is most sensitive to Einstein waves arriving from the zenith. Or from directly below—remember how Tony Tyson had to remind Joe Weber that the Earth is transparent to gravitational waves.

Now, if you want to measure the varying lengths of the two arms of the L, you can't use some kind of ruler. After all, it's spacetime itself that's growing and shrinking, so anything *in* spacetime will grow and shrink with it. If one arm of the L deforms, a ruler placed alongside the arm will deform in exactly the same way. Instead, scientists measure changes in length by changes in the amount of time that a light ray needs to travel from one end of the arm to the other.

One of general relativity's basic assumptions is the constancy of the speed of light. No matter what happens to spacetime, light will always travel at the same velocity: 300,000 kilometers per second. So if spacetime is stretched in a particular direction—if there's a little bit more space between two points—it will take light a fraction of a second longer to go from point A to point B. That means we can use clocks instead of rulers.

Physicists and astronomers are very familiar with accurate timing measurements. A good example was given in Chapter 6. The pulse arrival times of a binary pulsar are determined to a precision of better than a millionth of a second. That's enough to derive the masses and orbital characteristics of the system. As we've seen, it's even enough to detect gravitational waves, albeit indirectly.

Alas, sending pulses of radiation from one end of the arm and measuring their arrival times at the other end is way too inaccurate for our goal. Suppose we could measure the travel time of a pulse of radiation to a precision of one ten-millionth of a second (0.1 microsecond). That would enable us to measure changes in distance of about 30 meters (one ten-millionth of 300,000 kilometers). But we don't expect Einstein waves to arrive on Earth with such a large amplitude. (In fact, our bodies wouldn't survive such a dramatic stretching and squeezing of spacetime.) So using radiation pulses won't work either.

If submicrosecond precision is insufficient to detect gravitational waves on the ground, you may wonder how Joe Taylor and Joel Weisberg managed to prove the existence of the waves using pulsar timing. The answer, of course, is that they could wait for the effect of orbital decay to build up over the years. With LIGO, that wouldn't work: waves have to be detected at the very moment they pass through the detector. The only solution is to dramatically increase the sensitivity. We need to be able to measure variations in travel time to a precision of something like a billionth of a billionth of a second. No clock is that precise.

The solution is known as interferometry. It's what the *I* in LIGO stands for. And it all has to do with the wave nature of light. Interference can be seen in a pond. If you throw a stone in the water, it will create a concentric pattern of waves. If you throw a second stone a few meters from the first, it, too, will produce waves. The two sets of waves will now interfere with each other. At some points, where two wave crests arrive at the same time, the waves will add up to make a higher wave. At other points, where the crests from one set of waves meet up with the troughs of the other, the waves will cancel each other out. The result is an interference pattern of doubled and suppressed water waves.

With light, it's the same. If two light waves are in phase, with matching crests and troughs, they will reinforce each other. In other words, the amplitude will double (and the energy will increase). That's called constructive interference. If, however, the light waves are out of phase, with the crests of one wave matching up with the troughs of the other, they will extinguish each other. That's destructive interference.

Now suppose we have two rays of orange light with a wavelength of 600 nanometers (0.6 micrometer). They start out exactly in phase but in different directions. After traveling for a while, both rays hit a mirror and are reflected back to where they came from. If the two mirrors are placed at precisely the same distance from the light source, the waves will still be in phase when they meet up with each other again. As a result, the combined light is brighter than each individual beam.

But now let's assume the distance to one of the two mirrors increases a tiny bit—such a small amount that the light's travel time is extended by one femtosecond. A femtosecond is a millionth of a billionth (10^{-15}) of a second. In one femtosecond, light covers a distance of 300 nanometers. So upon arrival at the point of origin, one light wave now lags behind the other by half a wavelength. The crests and troughs of the two light waves no longer match. Instead, they're now out of phase (in this particular case they're in antiphase, with the crests of one exactly meeting the troughs of the other). The result is that the waves cancel each other out.

So interferometry lets you detect travel time differences on the order of femtoseconds. Maybe that's not yet precise enough for our needs, but we're getting somewhere.

For this purpose, it's easiest to work with light at one particular wavelength (or color). White light contains all the colors of the rainbow. With its whole range of different wavelengths, white light isn't well suited for interferometry. But laser light has only one very specific color—only one particular wavelength. So we'll definitely need a laser. That's what the *L* in LIGO stands for. Incidentally, LIGO uses not visible-light lasers but near-infrared lasers with a relatively long wavelength of 1,064 nanometers.

How do we get two beams of laser light to be exactly in phase? That's the easy part: just use one beam and split it in two using a beam splitter. A beam splitter is a mirror that reflects only half of the incident light. The other half of the light is going straight through the mirror. In fact, your sunglasses are a good example of beam splitters. Part of the incident light passes right through them (otherwise you wouldn't see a thing); the remainder is reflected back. Needless to say, LIGO's beam splitters are much more sophisticated than your average pair of sunglasses.

Laser, beam splitter, mirrors, detector. That's the basic design of LIGO—and of all other gravitational wave interferometers. (Yes, there are more; I'll come back to that in Chapter 8.) The laser produces a beam of monochromatic light. Let's assume the laser beam is aimed due east (in the diagram to follow, it travels from left to right). The beam splitter is oriented diagonally with respect to the laser beam. One-half of the laser light travels right through the beam splitter into the east arm of the L-shaped observatory. The other half is reflected to the side ("upward" in the diagram) into the north arm of the L, which is perpendicular to the first one.

At the end of each arm is a mirror reflecting the infrared light back toward the beam splitter. Again, half of the returning light passes through; the other half is reflected. So now we end up with light waves traveling westward (to the left in the diagram), back in the direction of the laser, and light waves traveling southward (to the bottom), in the direction of a photo detector—a sensitive light meter that converts light into an electrical signal. Because of all the optics and reflections involved, it just so happens that addition (constructive interference) will occur only for the westward-traveling light waves. The light waves that end up traveling southward, in the direc-

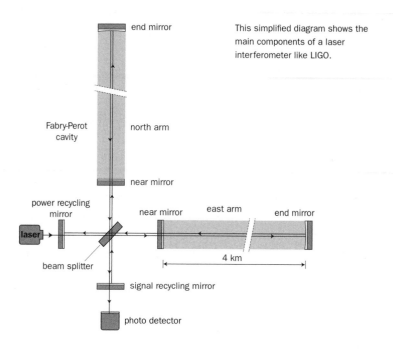

end mirror

This simplified diagram shows the main components of a laser interferometer like LIGO.

Fabry-Perot cavity

north arm

near mirror

power recycling mirror

near mirror

east arm

end mirror

laser

beam splitter

4 km

signal recycling mirror

photo detector

tion of the photo detector, cancel each other out—destructive interference.

The thing is, light cannot just disappear when two light waves are out of phase. If there's destructive interference in one direction, there must be constructive interference in another direction. Conservation of energy is one of nature's laws that we can't violate. (In the pond, it's the same: the water waves from the two stones cancel each other out at some locations. But that's possible only because they're adding up elsewhere.) So when there are equal arm lengths—the default situation—laser light exits the interferometer in the direction it came from, while the photo detector sees nothing. That's why the south side of the beam splitter is called the dark port.

But what happens if a gravitational wave passes through? Well, then the arm lengths (and the corresponding light travel times) will vary. First the north arm will grow while the east arm shrinks; next the north arm will get squeezed while the east arm gets stretched out. Light returning from one end mirror takes an infinitesimal fraction of a second longer to arrive at the beam splitter than does light returning from the mirror in the other arm. This causes a subtle change in the resulting interference. Constructive interference in the direction of the laser isn't 100 percent perfect anymore. Neither is the destructive interference in the direction of the photo detector. Even when there's an incredibly tiny length difference (much smaller than the wavelength of the laser), *some* light will start to trickle into the dark port. The sensitive photo detector should be able to pick this up. Bingo—we've detected a gravitational wave.

I've explained how interferometry lets you detect tiny differences in light travel time between two coherent laser beams. Obviously, it helps to make the arms of your interferometer as long as possible. A passing Einstein wave will stretch and squeeze spacetime by a certain degree—a certain percentage. For instance, the distance between two points may grow and shrink by no more than one-quintillionth of a percent or so (one part in 10^{20}). For two points close to each other, this amounts to almost nothing. The resulting variations in light travel time are so tiny that they're undetectable. But for two points at a greater distance, the light travel time variations will increase accordingly. So the longer the arms of the L, the easier it will be to detect gravitational waves of a certain amplitude.

Isn't 4 kilometers long enough? Not really—1,200 kilometers would be better. Now go tell that to your funding agency. But there's a smart solution, as always. Just fool the laser beam into thinking that it's traveling along a 1,200-kilometer tunnel. How? By using two mirrors in each arm instead of one. The first mirror is the one at the far end of the arm. The second one is at the near end of the arm, close to the beam splitter. If the laser light bounces up and down between the two mirrors a few hundred times, you've effectively created a 1,200-kilometer arm. The light travel time also goes up by a factor of 300 to a few milliseconds. Tiny variations on the order of one part in 10^{20} are now easier to detect.

After a few hundred bounces, the light has to come out of its temporary "prison," of course. That's easy. If the mirror at the near end of the arm reflects 97 percent of the incident light, the remaining 3 percent will pass through the mirror and exit at the other side. This is another way of saying that, on average, any photon of light will be reflected about 300 times before it escapes the prison. (Our 4-kilometer-long light prison is officially known as a Fabry-Perot cavity.)

But the escaping light still has to be a coherent laser beam or it won't be able to interfere with the beam that's coming out of the other arm. So while the light bounces up and down between the two mirrors, it has to stay in phase with itself. The only way to achieve that is to keep the round-trip between the two mirrors equal to a whole number of wavelengths. That calls for picometer precision (one picometer is 10^{-12} meters, or a billionth of a millimeter). Any departure destroys the final interference pattern. As LIGO scientists say, the interferometer arm needs to be locked.

The way to do this is by using an intricate feedback mechanism. If the round-trip between the mirrors is kept constant at a whole number of wavelengths, the photo detector at the dark port of the interferometer will register nothing. But if the length of the arm changes because of some external vibration, some light will start to trickle into the detector. As soon as that happens, a signal is sent to a control device at the end mirror of the arm. An electrical current is sent through a coil, creating a magnetic field. Small magnets on the rim of the end mirror start to feel an attractive or repulsive force. Apart from magnets, LIGO also uses electrostatic pushers, which use the same force that attract snippets of paper to an electrostatically charged pocket comb. Thus, the mirror can be moved back and forth a bit—enough to restore the lock of the arm.

Of course, a passing gravitational wave will also destroy the original interference setup because of the resulting variations in light travel time. Again, the photo detector will start to register some light. The feedback mechanism will respond, varying the electrical current through the coil and changing the strength of the magnetic field. As a result, the mirrors will be moved in such a way as to restore perfect destructive interference at the dark port.

So if you continuously read out the varying electrical current that is sent through the coils, you have a clear picture of the tiny, forced motions of the mirror over time. Most of this lock-restoring motion is necessary because of external vibrations ("noise"), but some of it may be due to a genuine Einstein wave.

Temporarily storing laser light in the interferometer by using two mirrors in each arm has the additional advantage of building up power in the two arms. As a result, the light that exits the Fabry-Perot cavity is a much more powerful and steady flow of photons than the light that went in. That's important if we need to measure extremely small changes in the output, which we do.

To understand why more photons enable a higher measurement precision, let's imagine you want to know exactly how hard it is raining during a summer storm in Louisiana. You happen to be in a shed with a metal roof, and all you've got is an old-fashioned sound meter, where the needle moves through an arc. So you decide to use the sound on the roof as a measure of the rain's intensity. If the rainfall is light, you will just hear *tap . . . tap . . . tap-tap . . . tap*. It will be very hard to say just how loud the rain is, and the needle on your sound meter will bounce around wildly. This effect is known as "shot noise." But then the storm rolls in, and the rain really starts to pour down. The needle moves up on the scale and settles to a steady value, which you can read off with high precision. That's why we want lots of light—lots of photon "raindrops"—to know just how much the light level changes as the mirrors move.

So now we've created an almost-ideal interferometer. It has virtual arms of about 1,200 kilometers, enabling the detection of extremely small variations in light travel time. If such variations occur, the dark port will no longer be completely dark. Instead, some light will enter our photo detector. By pumping up the laser power in the two interferometer arms, we have greatly reduced shot noise. The result is that even the tiniest changes in the amount of light due to a passing Einstein wave clearly stand out against the remaining noise.

———

Of course, there are many additional problems that hamper the search for gravitational waves.

By far the biggest problem, as you can imagine, is external vibrations: the slamming of a door or the passing of a truck, people walking in the neighborhood, industrial activity in a nearby town, tiny temperature changes, distant thunder, impacting air molecules, logging activity in the forest next door (in the case of LIGO Livingston), Pacific Ocean waves crashing on the coast of southern Washington State (in the case of LIGO Hanford), seismic tremors—the list goes on and on. The mirrors have to be isolated to the highest possible degree from all this "seismic noise," as it is called. Otherwise, you will never be able to recognize the tiny effect of a passing gravitational wave.

That's why so much effort has gone into designing the intricate suspension systems of the mirrors. To isolate the mirrors from external motions, almost every available technique has been applied. Vibrational sensors provide input for active damping systems that counteract ground motions, more or less the same way noise-canceling headphones work. Further isolation is provided by complicated sets of cantilevered flat blade springs and cushions. But the most efficient technique is the application of the pendulum mechanism.

A very simple experiment will reveal the damping power of a pendulum. Take a piece of thin rope or kite string about a meter long. Tie it to the handle of a heavy coffee mug. Grab the end of the string and let the coffee mug hang still. If you now slowly move the top of the rope to the left and to the right, the coffee mug will duly follow this motion. But if you move the rope much faster, the mug hardly moves at all. It works even better if you tie a second coffee mug below the first one using a second piece of rope: fast motions of the top end of the rope don't seem to affect the lower mug at all. Likewise, a suspended mirror can be isolated from high-frequency vibrations from its surroundings. LIGO uses a four-stage pendulum. It helps that the mirrors are thick and heavy—LIGO's mirrors measure 34 centimeters across, have a thickness of 20 centimeters, and weigh in at some 40 kilograms. It helps that they are suspended by the thinnest possible wires (0.4 millimeter) and made of fused silica—special glass that is very strong. It helps that the mirrors have an extremely high level of purity and simplicity—in the case of LIGO, they're simply very highly polished cylinders made of fused silica.

Obviously, it's impossible to get rid of *all* vibrations. There will always be at least *some* remaining seismic noise, some residual motions of the mirrors, however tiny. To make absolutely sure that an incredibly weak gravitational wave signal is recognized for what it really is, it's necessary to have at least two identical detectors separated by hundreds or even thousands of kilometers. The background noise will be different for the two observatories, but any cosmic gravitational wave signal will be similar. There may be subtle differences in the details depending on the direction of origin and the relative orientation of the two interferometers. But the facilities at Livingston and Hanford should both detect the same gravitational wave within a time span of a hundredth of a second. (In fact, between 2002 and 2010, a passing gravitational wave would have been detected by *three* instruments. Not many people know that the Hanford facility originally contained two separate, completely independent interferometers: one with 4-kilometer arms and one with arms half as long, both housed in the same tunnels.)

Needless to say, the laser, the beam splitter, and the photo detector also have to be isolated from external vibrations as much as possible. Moreover, all sensitive parts of the interferometer are encapsulated in giant vacuum tanks. Even the 4-kilometer arms—the steel pipes through which the laser beams bounce up and down—are completely evacuated. You don't want the mirrors to jitter as a result of being bombarded by air molecules. You also don't want the laser light to get scattered, which is what air molecules and tiny dust particles do. At some 9,000 cubic meters, LIGO's high-vacuum system is one of the largest in the world.

Yet another potential problem is the radiation pressure exerted by the laser beams on the mirror. Then there's "thermal noise"—extremely small molecular motions in the mirror coating caused by the ambient temperature of the whole experiment. And of course the slightly curved mirror surfaces should be polished to the highest possible degree, lest tiny irregularities destroy the coherence of the laser beam.

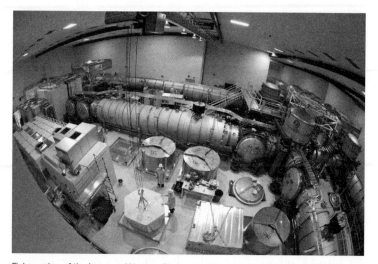

Fisheye view of the Laser and Vacuum Equipment Area (LVEA) at the LIGO Hanford Observatory.

The list of potential noise sources goes on and on—I've only begun to scratch the surface. All those effects tend to thwart the detection of gravitational waves. But for each and every problem, scientists and engineers have found a solution or a workaround.

Additional subsystems of the interferometer further increase the sensitivity. For instance, a laser cleaner (officially known as the input mode cleaner) makes sure that the laser light is as pure and stable as possible. The waves that go into the tunnels really have to have *exactly* the same wavelength and be precisely coherent.

Another indispensable element is the power recycling mirror. You may remember what happens when the returning laser beams from the two arms of the L meet again at the beam splitter: they will cancel each other out in one direction (the direction of the dark port) and reinforce each other in the other direction (the direction of the laser). So during normal operation, quite a lot of laser light is flowing back in the direction it came from in the first place. Not using all this laser power one way or another would be a waste of resources. So a power recycling mirror sends the light back into the interferometer. The result is even more photons bouncing up and down the tunnels, and with more laser power comes better sensitivity.

The much smaller amount of light that does arrive at the instrument's dark port every now and then is also put to use. It, too, is reflected back into the interferometer arms. This rather novel process is known as signal recycling. Scientists even experiment with so-called squeezed light—a quantum optics trick in which Heisenberg's uncertainty principle is tweaked in our favor. Never mind if you don't understand this in detail; few physicists do. It's the result that counts: yet better precision.

Big science such as gravitational wave physics is no child's play. Joe Weber's resonant bar antennas were pretty advanced—in fact, one of Weber's original detectors is now on display at the entrance of the LIGO Hanford Observatory—but building a working Einstein wave interferometer is playing in a very different league altogether. Everything here is pushing the very limits of science and technology. Nd:YAG lasers, input mode cleaners, beam splitters, ultra-high-vacuum technology, supersmooth fused-silica mirrors, vibrationless suspension systems, power and signal recycling, sensitive photo detectors, incredibly precise metrology—all of them have to work together seamlessly, and everything has to perform flawlessly.

And it does, as evidenced by the detection of GW150914. Almost a century after Albert Einstein first proposed the existence of the elusive ripples in spacetime, physicists finally succeeded in actually detecting them directly. Talk about perseverance.

———

Back in the spring of 1998, LIGO Livingston site manager Gerry Stapfer told me he was convinced that gravitational waves would be detected soon after the interferometer went online in 2002. "You have to believe in something," he said. But in 2010, after many months-long observing runs, nothing had been found. Apparently, as most people had expected, the first version of LIGO still wasn't sensitive enough to bag even one convincing signal in eight years.

Almost seventeen years after my initial visit, in January 2015, Frederick Raab, then the head of LIGO Hanford, was equally optimistic. "People will be very surprised if nothing is found," he told me. By then, a complete new set of lasers, mirrors, suspension systems, and detectors had been installed in the existing buildings and tunnels. Scientists, engineers, and technicians were busy commissioning the new equipment; lock in of one of the interferometer arms had just been achieved for the first time. The second incarnation of the detector, Advanced LIGO (aLIGO), was designed to be more than ten times as sensitive as Initial LIGO (iLIGO) once fully tuned. As a result, it would be able to pick up signals from ten times farther away, monitoring a volume of space a thousand times as large. Raab's optimism appeared to be justified.

Still, I remember driving back through the desert in the direction of my Richland hotel thinking about the many early misconceptions and false starts in the quest for gravitational waves. The greatest minds on the planet had been discussing the topic for decades. Einstein himself never was completely convinced of their existence. No single search so far—and there were many—had experienced any success. And now a new generation of brilliant scientists had put their money on these giant, expensive laser interferometers—the most sensitive measuring devices ever constructed in human history.

But what if they were wrong? What if Einstein waves turned out not to exist after all?

That night I couldn't sleep. Out there, some scientists had been working on the technique for more than forty years. They had fought technological setbacks, political hurdles, budgetary issues, and personal quarrels. They had given everything to arrive at this point: the completion of giant laser machines that should finally detect ripples in the very fabric of the universe. What if it all had been for nothing?

But as I lay there worrying, a propagating spacetime disturbance caused by two colliding black holes in the distant universe had almost completed its 1.3-billion-year trip to our tiny home planet in the outskirts of a galaxy called the Milky Way. Having lost its original loudness, it was now little more than a faint whisper, unnoticeable to all but the most sensitive ears. The first Einstein wave ever to be directly detected on Earth had already passed Proxima Centauri, our nearest stellar neighbor. It had stretched and squeezed frozen comets on one side of the Oort Cloud ever so slightly. Two-thirds of a light-year ahead was the Sun, orbited by a tiny blue planet.

Just in time, LIGO would be ready.

(((8)))

The Path to Perfection

The first time I met Rai Weiss was in an elevator in the Washington State Convention Center in Seattle. It was early January 2015. I was attending the 225th meeting of the American Astronomical Society. Weiss was about to give a presentation on the history of gravitational wave physics. We went down together for three floors or so. We said hi and bye but didn't exchange any conversation. "Quiet old man," I thought.

I was wrong. Yes, Weiss was eighty-two years old at the time, but he was certainly not quiet, as I found out later that day. After his presentation, I wanted to ask him a couple of questions, and he just didn't stop talking. Names, dates, events, suggestions for my book, technical details, nerdy jokes, personal stories—he was an avalanche of information. During an interview in the summer of 2016, it was exactly the same. I had requested forty-five minutes of his time. Instead, we talked for almost an hour and a half. Or, I should say, *he* did.

If anyone has great tales to tell about the history of LIGO, it's Rai Weiss. He's generally seen as the project's founding father, if not the inventor of the laser interferometry technique. He's an inspiring personality, too. Engaged, driven, empathetic—everyone who has ever worked with him has fond memories (well, *almost* everyone). And many believe LIGO never would have been realized without his brilliance and relentless enthusiasm.

Rainer Weiss was born in Berlin in the fall of 1932, just weeks before Albert Einstein left the German capital for good. As a young child he lived in Prague for a while; when he was seven, just before World War II broke out, the family moved to New York City. (Weiss's father was a Jewish physician.) Young Rai was a gifted and curious child, a tinkerer. He would repair the toaster. He would take a watch apart and put it back together again. He would stroll sidewalks in search of discarded electronic parts that might come in handy one way or another. As a teenager, he even started his own little business fixing broken radios and phonographs for his high school classmates.

By the late 1940s, Weiss had become an audio engineer of sorts. People hired him to build semiprofessional hi-fi systems. He never got terribly rich, but he made a fair bit of money. Why go to college? Well, he recalls, he wanted to learn more about noise-reduction techniques. The 78 rpm phonograph records that were popular those days produced a lot of crackle and hiss, and Weiss had not been able to find a solution to the problem. Studying electrical engineering at the renowned Massachusetts Institute of Technology in Cambridge would certainly help, he thought.

That turned out to be a bit too optimistic. Weiss's engineering classes were boring—he didn't really learn anything new. Would physics be more interesting? To a certain degree, yes. But once again, he got too distracted by other things in life to make much progress—like hopelessly falling in love with a beautiful piano player. "I followed her all the way to Chicago," he recounts, "but she must have felt I was too much in love to be useful to her. Eventually I went back to MIT."

Around 1960, the physics bug finally got him. The experimental physics bug, to be precise. As a graduate student of MIT's Jerrold Zacharias, he worked on building some of the first commercial atomic clocks. Those were the forerunners of "Mr. Clock"—the type of device that was taken on flights around the world some ten years later by Joseph Hafele and Richard Keating. In fact, Zacharias himself had plans to take his clocks to Jungfraujoch, a 3,470-meter mountain in the Swiss Alps. His goal: to measure the effect of gravitational redshift much more precisely than Robert Pound and Glen Rebka had just done at nearby Harvard.

The Switzerland experiment never materialized. Still, Weiss got hooked on everything that had to do with gravity and precision measurements, which turned out to be the right mix for someone who later would become one of the initiators of LIGO. For two years he did postdoctoral research with the famous physicist Robert Dicke at Princeton University building gravimeters. After returning to MIT, he set up a new research group on cosmology and gravitation. Cosmology is the science of the universe at large—you'll read more about it in Chapter 9. In the 1960s, the field was very much still in development. The big bang theory became ever more popular. In particular, 1964 saw the discovery of the cosmic background radiation often described as the afterglow of creation. To a physicist, it was clear that cosmology and general relativity were two sides of the same coin.

It's no surprise, then, that the staff of the MIT Physics Department asked Weiss to start teaching a course in general relativity. That was in 1967, around the time Jocelyn Bell discovered the first pulsar. But Rai Weiss was a tinkerer, not a theoretician. "The math was way beyond me," he says. "But I couldn't tell them I didn't master the topic, of course. It was one hell of a year. I spent all my free time learning about relativity. Sometimes I was just one day ahead of my students. They were much smarter than me."

Meanwhile, a few hundred kilometers farther southwest at the University of Maryland, Joe Weber was experimenting with his resonant bar detectors (you read all about it in Chapter 4). Hearing about this, Weiss's students got curious and asked him about the detection of gravitational waves. Again, mysterious stuff! But he found a neat way to explain the concept using three widely separated, free-floating "test masses" and accurate clocks—he knew everything about clocks. Don't imagine measuring changes in distance, he told his students. Instead, imagine measuring changes in light travel time. By now, that should sound familiar to you.

What Weiss did not know was that the concept was not entirely new. Two Russian researchers, Mikhail Gertsenshtein and Vladislav Pustovoit, had published similar ideas a few years earlier. But they had published in a Soviet journal that hardly anyone in the United States had ever heard of. One of the few American physicists who was in close contact with Soviet colleagues at the time was theorist Kip Thorne at Caltech in Pasadena. During the height of the Cold War, Thorne regularly spent time at Moscow State University to work with the precision-measurement group of Vladimir Braginsky; that's how he eventually learned about the earlier publications.

Anyway, Weiss worked out the fundamentals of a gravitational wave interferometer in a landmark paper that appeared in 1972 in MIT's *Quarterly Progress Report.* Almost forty-five years later, scientists like Thorne are still impressed by the publication. It contains most of the basic elements of the design. It also describes in detail the many noise sources that experimenters would need to deal with and, more importantly, how they might be able to do so. It certainly helped to sharpen the minds of the people who were already building the very first small prototypes.

So why didn't Weiss build a prototype detector following his own 1972 recipe? In fact, he did, but it took a while because of money issues. Originally, the physics department at MIT was largely funded by the Department of Defense. After World War II, the military needed as many new brilliant scientists and engineers as it could get. "We don't care what they do, but make sure people graduate" was the general message. But in the early 1970s, during the "insane" Vietnam War, as Weiss describes it, many left-leaning people felt uncomfortable with the situation. They felt the military shouldn't have any influence on the development of science. New legislation guaranteed that, in the future, the Department of Defense could support only science that was relevant to matters of national security. Cosmology and gravitation were not particularly relevant to matters of national defense, so Weiss lost his military funding, and MIT had few means to replace it as well as little interest in doing so. Before long, the institute's administration decided to dismantle the group. Weiss's work on a space mission to study the cosmic microwave background still received support from NASA, but his gravitational wave program was killed almost overnight. (The space mission eventually turned into the Cosmic Background Explorer satellite, or COBE.) He had to turn to the National Science Foundation (NSF) for a grant.

At the time, the NSF was still supporting Joe Weber's resonant bar experiments. What about this new interferometry technique? Was it really more promising? In 1974, the NSF sent out Weiss's grant proposal to various research groups for external review. "My ideas went all over the world before I got any money," says Weiss. It wasn't until the late 1970s that he finally received NSF money to start building his own small prototype interferometer.

An earlier prototype that was inspired by Weiss's ideas was a 3-meter interferometer in Munich, Germany. It was constructed by the gravitational wave group of computer pioneer and physicist Heinz Billing of the Max Planck Institute for Astrophysics. Billing was already building sensitive bar detectors to check on Joe Weber's claims. Like everyone else, he found nothing. But that didn't necessarily mean that Einstein waves did not exist, of course. Interferometry, about which Billing had learned by reviewing Weiss's NSF proposal, might be a more promising road to an actual detection. Why not give it a try? Another early prototype was a 2-meter tabletop experiment at the Hughes Research Laboratories in Malibu, California. This one was the brainchild of Joe Weber's former postdoc Bob Forward.

By the time Weiss got his own prototype up and running, Kip Thorne at Caltech had also helped launch an experimental group. Thorne himself was a pure theorist—a thinker, not a tinkerer. In 1973, together with Charles Misner and his former mentor John Archibald Wheeler, Thorne had written a 1,300-page textbook on gravity simply called *Gravitation*. Every physicist I have interviewed over the past couple of years has a copy of the fat black tome on the shelf. It's the bible of general relativity.

But Thorne had little experience with experiments. To Weiss's dismay, the original edition of *Gravitation* bluntly stated that laser interferometers would never be sensitive enough to actually detect

Einstein waves. The guy clearly had to be educated. That happened on a memorable night in 1975, in a hotel room in downtown Washington, DC.

Earlier that year, NASA had asked Weiss to chair a committee on space applications for gravitational physics. By then, the space agency was already involved in Francis Everitt's Gravity Probe B experiment, which you read about in Chapter 3. Weiss invited Thorne to share his views with the committee. "I picked him up from the airport," Weiss recalls. "We had never met. He hadn't booked a hotel. We shared a room. And we talked gravitational waves and experiments all night, until four or so."

They were two very different people. Weiss, forty-two at the time, looked like your average physics professor: sweater and sturdy shoes, or maybe a cheap tweed jacket and tie. Thorne was a thirty-five-year-old ex-hippie from California with long hair, a beard, an earring, and sandals. But they got along extremely well. "That night," says Weiss, "he completely turned around on the promise of laser interferometry. He's very smart."

Kip Thorne started to work out the expected number of detections for laser interferometers of various sensitivities. How often would they actually "feel" something? The most promising sources of spacetime ripples would be the violent mergers of neutron stars or black holes. At Arecibo, Russell Hulse had just discovered the first binary pulsar. It wouldn't be long before Joe Taylor and Joel Weisberg confirmed that the system was losing energy in the form of gravitational waves. Right now, the waves from this source are much too weak to be detected here on Earth. But they will become stronger over time, and when the two neutron stars eventually collide and merge, general relativity predicts a powerful burst of Einstein waves. For merging black holes, the expected amplitude would be even larger.

Collisions of neutron stars and black holes are extremely rare events in the universe. If such a catastrophe played itself out in our own Milky Way galaxy, even a simple Weber bar might detect the resulting gravitational wave signal. Unfortunately, this just doesn't happen often enough—it could take many thousands of years before the next one occurs. But a sensitive interferometer would be able to pick up the bursts of Einstein waves from mergers in other galaxies, out to tens of millions of light-years. Make your detector sensitive enough, and you may witness a handful of events per year.

Thorne also wanted to talk Caltech into funding real experiments—not just theory, but hands-on science that involved building a prototype and gaining experience. He wanted to succeed where Joe Weber had failed. Sure, identifying new opportunities and taking up intimidating challenges are what science is all about. But according to Virginia Trimble, Weber's widow, personal feelings may have played a minor role, too. "In the late 1960s, I had a relationship with Kip," she told me. "When Joe married me in 1972, Kip may have felt that Joe had stolen his ex-girlfriend."

Anyway, the Caltech group got started. Thorne would have loved to bring his Soviet friend Vladimir Braginsky to Pasadena. Braginsky was a real experimentalist, and Thorne had worked with him since 1968. However, that turned out to be a bridge too far given the political realities of the Cold War. Instead, following suggestions by both Braginsky and Weiss, Thorne approached Ron Drever at the University of Glasgow. With little money but great ingenuity, Drever had also built bar detectors. He was dabbling in laser interferometry, too, working on his own prototype instrument. He had one of the most creative minds in the field, always full of new and clever ideas. Starting in 1979, Drever divided his time between Glasgow and Pasadena. He took a position at Caltech as a faculty member in 1984.

So at the start of the 1980s, the focus in gravitational-wave physics was very much on laser interferometry. In Glasgow, an instrument with arm lengths of 10 meters was under construction. Larger would have been better, but the thing had to fit in the university's physics laboratory. In Munich, Heinz Billing and his colleagues had built a sensitive 30-meter prototype. Its size was determined by the dimensions of the garden at the Max Planck Institute for Astrophysics. At the northeast corner of the Caltech campus, a warehouse-like building became the home of the 40-meter prototype that was Ron Drever's new toy. Again, the scale of the detector was limited by the available space.

Meanwhile, in Cambridge, Massachusetts, Rai Weiss and his team of graduate students and postdocs had to content themselves with a tabletop instrument. Its arm length was a mere 1.5 meters—the largest they could afford, given their modest NSF grant. While Caltech invested some $3 million in the endeavor there, the MIT administration had not the slightest interest in supporting the new technology, says Weiss. "They thought laser interferometers would never be able to detect gravitational waves. Some high-ranking officials even were critical about general relativity and the existence of neutron stars and black holes. It has changed a lot since the 1990s, but back then, the atmosphere wasn't very intellectual."

None of this held Weiss back from working out the costs of an ambitious, 10-kilometer Long Baseline Gravitational Wave Antenna System. Weiss wrote up the proposal together with his MIT collaborators Peter Saulson and Paul Linsay, and with Caltech's Stan Whitcomb. The proposal, which became known as the "Blue Book," was intended to convince the National Science Foundation to fund a major R&D project.

Thanks to the enthusiasm of Richard Isaacson, NSF's program officer for gravitational wave physics, the 1983 Blue Book was taken very seriously and got favorable peer reviews. In due time, the development plan was approved by the National Science Board, the science policy advisory body to the president and the US Congress. One year later, government money started to flow: the first tranche of a multiyear grant for the research and development of what ultimately became LIGO. There was one condition, however: the teams at MIT and Caltech would need to work on it together. Fine.

Well, not really. Ron Drever was very reluctant to work in close collaboration with Rai Weiss. Yes, he would leave rainy Scotland for sunny California, but he had expected to develop the big machine all by himself. Moreover, the two scientists had very different ideas about the best possible approach. Drever had little confidence in the design that Weiss had originally come up with.

Looking back at the birth of LIGO in the mid-1980s, Weiss can now hardly believe that the newborn survived at all. Together with Thorne and Drever, he tried to manage the unwieldy project as best as he could. The three became known as the "troika"—Soviet vocabulary *did* make it onto the Caltech campus, even if Braginsky himself did not. "After LIGO's first detections in 2015, we received a lot of acknowledgment and honor," says Weiss, "but in fact, that's a shame. We were pretty incompetent. None of us had any solid experience with running such a huge program. It was a very complicated time."

There were personal issues, too. Weiss and Drever didn't get along with each other very well. "Frankly, it was impossible to work with him," says Weiss. "His intuition steered him in every possible direction. One day he would have this idea. The next day he wanted to pursue something else altogether. Some of Ron's ideas were very clever, but others were absolutely abysmal. He couldn't make up his mind—couldn't make a single decision. He was a child in man's clothing. He wanted to go everywhere, so for a long time we got nowhere."

It was time for a critical review. That's what IBM's Dick Garwin suggested in 1985. You'll remember Garwin from his "bar fight" with Joe Weber at the Fifth Cambridge Conference on Relativity in 1974. He was a well-respected science consultant to the government. And he was pretty doubtful about the prospects of LIGO. Following Garwin's advice, NSF organized a blue-ribbon study week in Cambridge in November 1986. Everyone came, recalls Weiss: Nobel Prize–winning physicists, experimentalists, laser engineers, experts in high-precision mirror fabrication, metrologists. In the end—probably to Garwin's surprise—the committee embraced the whole thing, deciding the time was right to build a large laser interferometer to detect the elusive Einstein waves. In fact, LIGO should consist of two identical facilities separated by a few thousand kilometers. Only then would it be possible to confidently identify a real cosmic signal amid the remaining background noise.

Kip Thorne, Ron Drever, and Rochus Vogt *(left to right)* pose in front of the 40-meter prototype interferometer at the California Institute of Technology in Pasadena. This photo was taken in the late 1980s.

The time was also right to professionalize LIGO's management structure. In the summer of 1987, the Weiss-Thorne-Drever troika was supplanted by a single project director, Caltech's Rochus "Robbie" Vogt. The good thing about that decision was that Vogt got everything back on track. Decisions were made, deadlines were met, problems were solved. Two years after his appointment, Vogt had accomplished his main goal: to produce a final, detailed proposal for the construction of the Laser Interferometer Gravitational-Wave Observatory ready for NSF's okay. Magnificent.

The bad thing about the decision was that Robbie Vogt was a very difficult person to deal with. He got people organized by issuing orders—and anyone who did not follow his orders was shoved out of the way. "From what I had heard about him, I knew he would be the right guy to make the project work," says Weiss. "But I didn't realize how complicated he was. Someone at Caltech once told me, 'You'll never be the same person anymore after working in a project that's led by Robbie.' He was right."

An important part of the proposal was its two-phase approach. The first incarnation of LIGO (now called Initial LIGO or iLIGO) would be ready in the early years of the twenty-first century. It would have the best possible sensitivity that scientists and industry could achieve in the course of the 1990s. Einstein waves from neutron star mergers should be detectable out to a distance of at least 50 million light-years. There are thousands of galaxies within that volume of space. So, with a bit of luck, iLIGO might catch one or maybe even two neutron star mergers during its planned operational lifetime of about a decade. At least that was Kip Thorne's optimistic estimate.

But iLIGO would also be a proof of concept. Its main purpose was to get firsthand experience with all the novel technologies, find out where unexpected problems would arise, and demonstrate that running two large facilities in tandem really was doable. Meanwhile, the development of even more sensitive equipment would continue: more powerful, ultrapure lasers; higher-quality mirrors with higher-quality coatings; better suspension systems; and cleverer interferometer configurations. Advanced LIGO would go online around 2015. It would be the real thing, eventually reaching ten times the sensitivity of its predecessor, ten times the distance, and a thousand times the volume. Who knew, maybe it would yield dozens of detections per year.

When the LIGO proposal was submitted to the National Science Foundation in December 1989, 2015 was still a quarter century in the future. Those were bold plans. Fittingly, the document's preface began with a 1513 quote from Niccolò Machiavelli: "There is nothing more difficult to take in hand, more perilous to conduct, or more uncertain in its success, than to take the lead in the introduction of a new order of things."

Difficult, perilous, uncertain . . . but Machiavelli said nothing about expensive. Still, in 1990, the National Science Board approved the proposal despite the price tag of almost $300 million. There was just one caveat: because of the large amount of money involved—unheard of in NSF's history—the project had to get the green light from Congress, too. Capitol Hill would have the last word on the construction of LIGO.

It was yet another hurdle, and it almost got LIGO killed, partly because of Tony Tyson of AT&T Bell Laboratories. You'll remember Tyson—he was one of the strongest opponents of Joe Weber in the early 1970s. Tyson was asked to testify for the US House Committee on Science, Space, and Technology. His first assignment: to assess the feasibility of the LIGO endeavor. His second assignment: to carry out a survey within the astronomy community about LIGO's popularity.

Looking back at the episode, Tyson now wishes he had never accepted that second assignment. His own views on the LIGO proposal had already met with harsh criticism from the gravitational wave cohort. He was enthusiastic about the prospects, but he felt that the project was premature. Perhaps Congress should consider first spending money on intermediate-scale prototypes, rather than going all out right away. Those were technical considerations, in a sense. But Tyson's survey of some two hundred prominent US astronomers had a much more political ring to it. As it turned out, five out of six astronomers were against building LIGO at all. It was too difficult, too perilous, too uncertain, and above all, too expensive. Why not spend the money on new telescopes and astronomical instrumentation? On things that had proved their value?

Part of the reluctance had to do with the physics background of LIGO. A National Research Council advisory report on priorities for astronomy and astrophysics in the 1990s described LIGO as "an interesting physical experiment that has not shown to have astronomical relevance yet." And those physics guys with their laser facility had the guts to call their instrument an *observatory*? Nothing had been detected so far. You couldn't even point the thing at a particular position in the sky.

Of course, Tyson was obliged to report the outcome of his survey. And as a result, he got a slew of vitriolic emails from LIGO people. Nevertheless, Congress eventually approved the project thanks to two more years of intense lobbying by Robbie Vogt, who was a relative newcomer to Capitol Hill but whose eccentric personality got the attention of lawmakers. Finally, in 1992, twenty years after Rai Weiss first published his seminal paper in MIT's *Quarterly Progress Report*, the National Science Foundation was allowed to sign a cooperative agreement with Caltech and MIT. The two "observatory" sites—Hanford and Livingston—had also been picked. Construction of LIGO could finally begin.

Or could it?

Not really, at least not right away. Personal tensions at the Caltech campus would come to an embarrassing climax first. When Vogt had started out as LIGO's first director, he had relied on the technical expertise of Ron Drever, the man with brilliant ideas like boosting the laser power by using Fabry-Perot cavities and using power recycling to further decrease shot noise. But over the years, Vogt found it harder and harder to cope with Drever's infamous intuition and indecision. The huge change of scale of the project required structure, organization, and discipline—qualities that didn't appear to mean much to Drever.

Discussions turned into arguments; arguments turned into fights. Then the two men stopped talking altogether. Vogt would ostentatiously leave a meeting room as soon as Drever entered—not a good way to make progress on a multimillion-dollar project. Other team members also found it difficult to work with Drever. And many had problems with Vogt's rough and rigid management style. According to Weiss, it was an unbelievable mess. The science press got word of it; news stories about the "Vogt-Drever tempest," as it was called, appeared in the pages of *Science* and *Nature*. Eventually, after strong words from NSF, the Caltech faculty feared for the future of LIGO. In 1992, they threw Ron Drever out of the project. They even changed the locks on his office door.

But it seemed that too much damage had been done. An outside NSF review panel had even suggested canceling the project. Vogt became ever more suspicious of any form of outside interference because he wanted to run LIGO his way. That didn't go over well with the National Science Foundation, which requested a much more open mode of operation with a justification for every dollar spent and a proper report for every step taken. Could the NSF really trust Robbie Vogt? Sure, he had been instrumental in pushing LIGO through Congress.

But in the end, an outside review panel concluded he wasn't the right person to actually build the thing. NSF officials, together with the senior administration at MIT and Caltech, came to an unavoidable conclusion in late 1993: Vogt would have to leave, too. LIGO was too important to be ruined by the idiosyncrasies of one man.

So who *would* be the right person to finish the job?

Maybe Caltech particle physicist Barry Barish. He was an easygoing person, extraordinarily well organized. He had lots of experience in running large science projects. Until recently, he had been one of the coleaders of a big experiment intended to serve as a lead-up for the SSC, the Superconducting Super Collider. The SSC was a gigantic US particle smasher that would lead the hunt for the elusive Higgs boson. If the name of the facility is unfamiliar to you, that's probably because it was never built. In October 1993, the multibillion-dollar accelerator—funded through the Department of Energy—was canceled by Congress. So Barish had time on his hands, too.

Over the 1993 Christmas holidays, Barish was approached by Caltech president and fellow physicist Thomas Everhart. Would he think about taking the helm at LIGO? They discussed the request during a walk on the beach. Barish couldn't decide right away. He was still recovering from the SSC debacle. Moreover, he had been watching the LIGO project on and off, and he knew about all the fuss. Was this really a doable project?

In the end, Barish said yes. In February 1994, he replaced Vogt as principal investigator for the project. And using his skills as a strategist and manager, he succeeded in steering the LIGO ship into calmer waters, basically by doing away with the existing management structure, by bringing in a lot of new strength—many particle physicists were looking for a new job—and by formulating a much more realistic cost assessment of the project. He told NSF that if it really wanted to prepare for the development and implementation of Advanced LIGO some fifteen years in the future, it should expect the project to be some 40 percent more expensive than had been estimated before.

In the spring of 1994, when site preparations were about to commence at Hanford, two remarkable things happened. Yet another review panel now strongly recommended continuation of LIGO. And Barish, together with Kip Thorne, the project's chief theorist, was invited to testify at a meeting of the National Science Board in Washington, DC. He recalls it as a very formal event, lasting for an hour or so. Thorne described the science, including the best available expectations for the number of events that LIGO might detect. Barish described his new views on how to realize the project.

Looking back at that roller-coaster year, Barish calls it a "miracle" that the LIGO project was reapproved by the board in the summer of 1994, despite its much higher total cost. "But the bigger miracle," he says, "is that NSF has been funding this continuously for over twenty years so far. The potential payoff was high, but the risks were high, too. Then again, doing the best science you can always involves taking risks."

With NSF's greenlighting of the project, the Laser Interferometer Gravitational-Wave Observatory finally became a reality. Less than four years later, Livingston site manager Gerry Stapfer showed me around in his still-empty facility, beaming with confidence that a first detection was around the corner: "You have to believe in something."

———————

Via Edoardo Amaldi, in Santo Stefano a Macerata, is just a thirty-minute drive from the Piazza del Duomo in Pisa, Italy. In the historic city center, tourists take selfies in front of the Leaning Tower, maybe wondering why gravity hasn't yet toppled the edifice (basically because it's stabilized by a harness of steel cables). Some may even know the apocryphal story of Galileo dropping balls of different masses from the tower to prove Aristotle wrong. Few will realize that the most sensitive gravitational measurements on European soil are being carried out just half an hour to the southeast.

During my visit in the second half of September 2015, however, no measurements are being done at all, however, as the Virgo detector is being refurbished. "Advanced LIGO started a few days ago," says Federico Ferrini, director of the European Gravitational Observatory. "Like them, we are also installing new, more sensitive equipment. In

Aerial view of the Virgo gravitational wave detector near Pisa, Italy. Bridges allow farmers to cross the 3.5-kilometer beam pipes.

late 2016 or early 2017, we expect to join Advanced LIGO's second observing run." Right now, there are still many problems to solve and hurdles to jump. Big science is a matter of trial and error. A sign on the wall of Ferrini's office reads, "Let's make better mistakes tomorrow."

Only half jokingly, the Italian physicist tells me that a few weeks earlier, when he and his wife visited the Santuario di Montenero, a famous pilgrimage site close to Livorno, he had prayed for the detection of a real Einstein wave. "My term as director ends in late 2017," he says. "I'm sure that we'll have a few detections by that time." What he does *not* tell me is that the detection of GW150914 caused great excitement just eight days before my visit because no one in the LIGO-Virgo Collaboration is allowed to break the news yet. No wonder Ferrini sounds so confident.

Virgo is largely comparable to LIGO, although the length of its interferometer arms is 3 kilometers instead of 4. Also, the region southeast of Pisa is much more populated than either the Hanford Site in Washington State or the forests north of Livingston, Louisiana. Virgo's beam pipes lie above the ground, like their American counterparts. A number of low bridges had to be built to enable farmers to drive their tractors from one side to the other. The pipe covers have been painted sky blue so that they aren't too dissonant with the friendly Italian landscape.

Commissioning coordinator Bas Swinkels shows me around the facility. He is the only permanent Dutch scientist on the site—Virgo started out as a French-Italian project, but Hungary, Poland, and the Netherlands joined the collaboration at a later stage. Swinkels takes me into Virgo's Laser and Vacuum Equipment Area. It's huge but crowded with towering vacuum tanks. New to Advanced Virgo are the cryotraps, an important contribution from Nikhef, the Dutch National Institute for Subatomic Physics in Amsterdam. Using liquid nitrogen, they freeze out any remaining contamination in the system, resulting in a higher-quality vacuum. Swinkels also proudly describes Virgo's super-attenuators: 10-meter-high stacks of seven inverted pendulums from which the mirrors will be suspended on fused-silica wires.

Walking around the site, it's hard to believe that Virgo was little more than an idea until well into the 1980s—especially if you know how long it took LIGO to get started. Then again, the Europeans had the big advantage of not being the first. Extensive R&D had been carried out in the United States already.

Italian physicists are experienced gravitational wave hunters. In the early 1970s, Edoardo Amaldi and Guido Pizzella built their first sensitive bar detector with the goal of checking the claims of Joe Weber. Their group at the Frascati Laboratory of the National Institute for Nuclear Physics (INFN), near Rome, had been cooperating with

Heinz Billing's Max Planck team in Munich. They had found nothing convincing, but maybe laser interferometry would be the way to go.

At least that was the idea of particle physicist Adalberto Giazotto. Giazotto was an expert in seismic isolation. In the 1980s, he teamed up with Alain Brillet of the French National Center for Scientific Research (CNRS). Brillet knew all about optics and lasers. Together they came up with the first ideas for the Virgo detector—the European answer to LIGO. A formal INFN / CNRS project proposal was submitted to the French and Italian governments in 1989, just before Robbie Vogt completed the original LIGO proposal to the National Science Foundation.

The name Virgo is not an acronym like LIGO is; instead, the detector is named after the cluster of galaxies in the constellation Virgo. The Virgo cluster is 50 million light-years away; Giazotto and Brillet were aiming for a detector that would be able to pick up Einstein waves from merging neutron stars out to that distance.

With regard to general sensitivity, Virgo would be comparable to LIGO, even though the interferometer arms would be a bit shorter. But the Europeans wanted their detector to perform better at lower frequencies. How? Better mirror suspension. A huge multilevel pendulum system, designed by Giazotto, should do the trick. A working prototype had already been constructed in 1987, at the INFN lab in Pisa. It's now on display in the lobby of the European Gravitational Observatory's main building.

Virgo wasn't the only European initiative. In Germany in the late 1980s, there were also plans for a 3-kilometer interferometer a hundred times the size of Heinz Billing's 30-meter prototype. Billing had retired in 1989, but his pioneering work was continued by Karsten Danzmann. Billing, seventy-five at the time, had every confidence that the efforts of his successor would pay off eventually. "Herr Danzmann," he said, "I will live until you find these waves."

The Germans teamed up with experimentalists in Glasgow, Scotland, and with theorists in Cardiff, Wales. They called their future interferometer GEO, for German-English Observatory. That was stupid ignorance, Danzmann now admits: Scotland and Wales are part of the United Kingdom, but of course you should never call a Scots or Welsh person English. Before long, GEO came to stand for Gravitational European Observatory, although this full name is almost never used.

In the summer of 1990, it looked like the €100 million project was about to get the green light. However, over the next two years, GEO quietly dwindled away due to the fall of the Berlin Wall in 1989 and the subsequent reunification of East and West Germany. The lion's share of the new government's science funding went to restructuring efforts in the former German Democratic Republic. There simply was no money left for large new initiatives. By 1992, it was clear that GEO wasn't going anywhere, at least not in its original form.

New opportunities arose after Danzmann moved from Munich to Hannover, the capital of the German state of Niedersachsen (Lower Saxony). At the University of Hannover, renowned laser physicist Herbert Welling was restructuring the physics department, and experimental gravitational physics was high on his agenda. In 1993, he invited Danzmann to build up the new program. It would be funded partly by the Volkswagen Foundation—the German car company is headquartered in Niedersachsen. Before long, the GEO project was back on the table again, though much smaller and cheaper.

Virgo was approved in 1993, originally as a €75 million project. Construction work began three years later. The €10 million GEO600 project—the new name reflected the shorter arm length of just 600 meters—got started in 1994, with the first construction starting just south of Hannover in 1995. The Europeans were on a fast track.

A visit to GEO600 is a very different experience from a visit to LIGO or Virgo. First of all, the site is quite difficult to find. West of the tiny village of Ruthe, first look for the fields of the university's agriculture department. Then follow a narrow, muddy road to a loose collection of prefab buildings: the observatory's offices, control room, and canteen. The 600-meter corrugated beam pipes resemble the elements of a cheap wastewater system. They're hidden in trenches and are easy to miss. But appearances can be deceiving. When you walk into the central building, which is partly underground, you're suddenly surrounded by high-tech laser equipment, racks of electronics, and vacuum tanks containing precision optics.

At the time of my visit, in early February 2015, GEO600 was the only operational laser interferometer in the world—both LIGO and Virgo had been shut down to allow for the upgrade to the advanced detectors. No one expected the small German detector to really pick up spacetime ripples, though—it's much less sensitive than its three big siblings are. Instead, the main goal of the facility is to develop and test new technologies. Signal recycling was first pioneered here. GEO600 was also the first instrument to demonstrate the squeezed light technique, whereby quantum effects are used to further steady the interferometer's output.

At first, the European projects—especially Virgo—were seen as competitors to LIGO. Some thought the Europeans might even beat the Americans in making the first direct detection of Einstein waves; that remote possibility may even have played a role in the eventual survival of LIGO. But it soon became clear that everyone would benefit from a certain amount of cooperation.

Two years before LIGO's official inauguration ceremony in November 1999, GEO600 had already joined the LIGO Scientific Collaboration. The first coincident observing run of Hanford, Livingston, and GEO600 commenced in 2002. One year later, Virgo also became operational. And in 2007, the LIGO and Virgo collaborations signed a joint data-analysis agreement. Since then, all engineering data, test results, observational measurements, and scientific analyses of the four detectors are shared among the thousand-plus members of the various groups.

———

It had been a long and winding road, but all's well that ends well. After a long period of tuning, followed by several years of observation, the initial versions of LIGO and Virgo shut down in October 2010 and December 2011, respectively. Gravitational waves had still not been detected half a century after Joe Weber first started to think of ways to measure the tiny ripples. Yet everyone was optimistic. The construction of Advanced LIGO and Advanced Virgo was about to begin. In five years' time, the new detectors would be completed. Eventually they would reach much higher sensitivities than their predecessors had. Just a few more years, just a little bit more patience.

Then, on March 17, 2014, researchers at the Harvard-Smithsonian Center for Astrophysics in Cambridge, Massachusetts, announced what they described as "the first direct image of gravitational waves across the sky." Not from colliding neutron stars or from merging black holes, but from the big bang. And obtained not with a giant laser interferometer but with a small microwave telescope at the South Pole.

After decades of development, construction, and testing and after spending hundreds of millions of dollars, had Rai Weiss, Kip Thorne, Ron Drever, and all the others been scooped?

That's a story for Chapter 10. First, though, I have to explain the birth of the universe.

(((9)))

Creation Stories

"In the beginning there was nothing, which exploded."

This famous Terry Pratchett quote is often used (or misused?) to ridicule cosmology. The thread usually goes like this: You call yourself scientists? Claim to know things about the universe? Come on, this whole big bang idea must be a farce—it doesn't make sense at all. Which means science can't be the path to the truth. Bring back the divine Creator, or whatever.

I've never quite understood the argument. Science doesn't know how to cure cancer. Science has almost no idea about human consciousness. No one sees that as a reason to write off the scientific endeavor altogether. To the contrary, I would say. But here's the biggest, grandest, deepest, and most profound question of all—how did *everything* begin?—and scientists are laughed at because they haven't solved the mystery yet. What did you expect?

If you don't understand the origin of the universe, you're in excellent company. Even the smartest cosmologists don't know how it all began. The brightest minds in the world have no clue what happened before the big bang—or if that question even makes sense. Not even Stephen Hawking knows for sure whether the universe is truly infinite and whether there may be more than one. The biggest questions—the ones that every little child comes up with—have not been answered yet. Maybe they never will. But science has come a long way since the allegorical myths of ancient times.

If you ever tried to wrap your head around cosmological questions, I'm pretty sure you've run into trouble. Everyone does. Cosmic expansion, galaxy redshifts, curved space, infinity—these are tough topics. Cosmology is no walk in the park. But we've got a whole chapter ahead of us. I'll do my best to guide you through the conceptual minefield.

————————

Everyone has heard about the big bang. Some 13.8 billion years ago, all of the universe was compressed in one, dimensionless point in space, and the big bang exploded the stuff away in every possible direction, right?

Wrong.

There's your first—and biggest—misconception. The big bang was not an explosion *in* space. Instead, it was an explosion *of* space. At least, that's a much better way to put it. Most people imagine the big bang as a huge fireworks display: it goes off at a particular point and then hurls stuff through space in all directions. But as soon as you catch yourself thinking about the big bang as fireworks, stop thinking. It's the wrong image altogether.

To explain, let's jump back in time about a century. Astronomers had discovered spiral nebulae such as Andromeda and the Whirlpool. No one knew their true nature. Some people thought they were relatively close swirling gas clouds from which a new star might form eventually. Others thought they were large groupings of stars much farther away—well beyond our own Milky Way galaxy.

Measuring the distance to a spiral nebula was impossible. You can't just roll out your measuring tape from here to Andromeda. But there are a lot of other things about spiral nebulae you *can* figure out: sky position, apparent size, brightness, and shape. The more you know about them, the better your chances of discovering their true nature.

Vesto Slipher realized he could measure something else, too: a nebula's motion toward or away from us. Like his younger brother, Earl,

Vesto was an astronomer at Lowell Observatory in Flagstaff, Arizona. While Earl focused on planets, Vesto was more interested in nebulae. In 1912, he was the first to measure the velocity of a spiral nebula.

How can you measure an object's velocity if you don't even know its distance? By using the Doppler effect, which I described in Chapter 6. Just think of the passing ambulance again. When it turns onto your street on one end and races toward you, siren wailing, you perceive a high pitch. When the ambulance turns off your street on the other end, the sound of the siren is markedly lower. The change in pitch is a measure of the ambulance's velocity.

With light, it's the same. If a star moves toward us, the light waves we observe are compressed, so we see light with a higher frequency, corresponding to a slightly bluer color. If the same star moves away from us, we perceive a lower frequency or a redder color. From the observed tiny shift in color, you can derive the star's velocity, even if you don't know how far away it is.

By the early twentieth century, astronomers already had a lot of experience measuring these so-called radial velocities of stars (toward us or away from us, along the line of sight). But for a spiral nebula, it's much more difficult. A nebula is not a well-defined pinprick of light, like a star. Instead, it's a hazy smudge—and pretty faint, for that matter. Still, Slipher succeeded. Other astronomers at other observatories followed suit.

If you could measure the velocities of all ambulances in your neighborhood, you would expect to find about half of them more or less approaching you and the other half more or less receding from you. If not, you'd have to conclude that you're in a special position. At the site of a massive accident, for instance, more ambulances would race toward you (you'd hope). But at a random position, you would hear just as many high-pitched sirens as low-pitched sirens.

So you can imagine that Slipher and his colleagues were surprised to find that all the spiral nebulae they could observe were receding (except for one, as I'll explain later). In each and every case, the light observed from Earth had a lower frequency, corresponding to a redder color. In other words, all nebulae were redshifted. That was exceedingly weird—it made it seem that the Earth occupied some special position in the universe.

Before we move on, you should realize that this redshift is a very subtle effect. It's not that the Whirlpool Nebula has a ruddy hue. The frequency shift and the corresponding shift in wavelength (or color) are much too small to perceive with the eye. Instead, astronomers have to measure certain features in a nebula's light very precisely. For instance, hot hydrogen is known to emit red light at a wavelength of 656 nanometers (0.000656 millimeters). But in a given spiral nebula, this emission might be observed at, say, 658 nanometers. Still, this small shift would indicate a recession velocity of some 900 kilometers per second.

So here was the mystery: all spiral nebulae appeared to be receding, and at pretty high velocities, too. No one could think of an explanation—that is, until the late 1920s, thanks to American cosmologist Edwin Hubble. That name may ring a bell: the Hubble Space Telescope is named after him.

In 1924, Hubble had already proved that spiral nebulae are not part of our Milky Way galaxy. Instead, they are "island universes"—as astronomers used to call them back then—galaxies in their own right, containing billions of stars. Now, in 1929, Hubble made an astonishing discovery. The farther away a galaxy is, the faster it is receding from the Milky Way. Nearby galaxies move away from us at a moderate pace; distant galaxies have much higher velocities.

Of course, Hubble didn't know the distances to other galaxies very precisely. But he made educated guesses. If stars (or glowing gas clouds) appeared brighter in galaxy A than in galaxy B, it was safe to assume that galaxy B was farther away. Using those kind of guesstimates, the trend was obvious. Nearby galaxy: small recession velocity. More distant galaxy: larger recession velocity. Very distant galaxy: very large recession velocity.

In 1927, a Jesuit priest and astronomer in Belgium, Georges Lemaître, had already been the first to draw the right conclusion. Do we occupy a very special position in the universe? No. Are other galaxies mysteriously speeding away from the Milky Way? No. Are we really measuring recession velocities? No. Instead, space itself is expanding, in accordance with one particular solution of Albert Einstein's relativity equations. Quite rightly, Lemaître is considered the father of the big bang theory.

To explain what's happening, I'm going to use a raisin cake as an example. It's a well-known and often-used metaphor; I haven't even been able to find out who was the first one to come up with it. So here's your raisin cake thought experiment (well, you could of course really carry it out, but there's no absolute need, unless you're particularly fond of raisin cake).

Before putting the cake in the oven, we prepare it in a very special way: all raisins are positioned in the dough at a very regular spacing, one centimeter apart from their nearest neighbors. That's to say, the raisins are at the vertices of an imaginary lattice of cubes with each cube measuring 1 centimeter by 1 centimeter by 1 centimeter. Make sure that you have a clear mental picture of this setup.

Next, we fire up the oven. Since we're using super dough (it's a thought experiment, after all), the raisin cake will rise spectacularly. In fact, in just one hour of baking, the cake will end up at twice its original dimensions. So after one hour, each raisin is now 2 centimeters away from its nearest neighbors.

Now imagine you're sitting at one particular raisin. At first, your neighbor is 1 centimeter away. But after the baking, its distance is 2 centimeters. In one hour, its separation has grown from 1 to 2 centimeters. In other words, during your time in the oven, you see this neighboring raisin move away from you at a velocity of 1 centimeter per hour.

But the *next* raisin in line starts out at a distance of 2 centimeters and ends up at a separation of 4 centimeters. So that one appears to move at 2 centimeters per hour. Likewise, a more distant raisin, at 10 centimeters, also ends up twice as far—you see it move at 10 centimeters per hour.

Nearby raisin: small recession velocity. More distant raisin: larger recession velocity. Very distant raisin: very large recession velocity. Just what Hubble found.

There's an important thing to realize here. It doesn't matter at which raisin you're located. The view is the same from every point in the cake. No single raisin has a special position. Each and every one sees the others fleeing away. Likewise, our Milky Way galaxy doesn't occupy a special position in the universe. In every other galaxy—Andromeda, the Whirlpool, NGC 474—alien astronomers would observe exactly the same pattern.

The second important thing to realize is that the raisins aren't moving at all. At least not with respect to the dough. They're just sitting there. Sure, their mutual distances do increase. But that's not because they're moving; it's because the dough is expanding. Likewise, the galaxies in the universe are not racing through space at colossal speeds. Yes, the distances between them do increase, but that's because space itself is expanding.

I told you to be suspicious of a fireworks-like mental image of the expanding universe. In a fireworks explosion, bits and pieces of glowing material are moving through space away from their point of origin. They will end up farther away from each other than they were when they started, but that's because they actually moved. Not so in the expanding universe. You might say that the distance between Earth and the galaxy NGC 474 (at present some 100 million light-years) is growing at a rate of some 2,000 kilometers per second. But it would be wrong to say that NGC 474 is speeding through space at this velocity. Its distance is increasing because the space between the galaxy and us is expanding.

Admittedly, this is where the raisin cake analogy starts to break down a bit. Raisins do not move through dough. Galaxies, in contrast, *do* exhibit some real motion in space. For instance, our Milky Way galaxy and the neighboring Andromeda galaxy are actually closing in on each other at about 100 kilometers per second—that's the one exception that Vesto Slipher found back in 1912. It's just because the two galaxies feel each other's gravitational pull. They really are moving through space, heading for a collision a few billion years from now. (Don't panic—by then, life on Earth will have already been obliterated by the continuous swelling of the Sun, as I described in Chapter 5.)

And since their current distance is a mere 2.5 million light-years, there's not enough expanding space between the Milky Way and Andromeda to compensate for the mutual approach. On the other hand, the motion through space of a very remote galaxy is far too small to overcome the distance increase because of the expansion of all the space between the remote galaxy and the Milky Way.

There's yet another reason the raisin cake analogy isn't perfect. Raisin cakes are usually finite in size. In contrast, as I'll discuss shortly, the universe may very well be infinite. But that's nitpicking. For all practical purposes, the cake metaphor works beautifully. In any case, next time you're imagining the expanding universe as fireworks, the alarms should go off, and you should shout to yourself, "Raisin cake, raisin cake!"

Now what about a galaxy's redshift? Wasn't that due to the galaxy's motion away from us? Sure—the motion in space of a galaxy produces what appears to be a Doppler effect. If it's moving away from us, its light waves are stretched to lower frequencies and corresponding longer wavelengths, resulting in a redshift. If it's moving toward us (like Andromeda and just a handful of other small nearby galaxies), it exhibits a small blueshift. But if we're talking expanding space, it's better to forget about the ambulance analogy.

Instead, try to imagine a light wave emitted by a distant galaxy with a certain frequency and a corresponding wavelength. For millions or even billions of years, the light wave travels through space on its way to a telescope here on Earth. If we lived in a static universe, it would arrive at Earth with exactly the same wavelength as it started out with. But the universe is not static. Space is expanding. As a result, a light wave traveling *through* space is expanding as well. It is slowly stretched to a longer wavelength—a redder color.

The longer a light wave travels through expanding space, the more it will be stretched. So the light from distant galaxies—longer travel times—will be redshifted to a larger degree than the light from nearby galaxies. Again, just what Hubble found. In fact, cosmologists use a galaxy's redshift as a proxy for its distance.

Now we're getting somewhere. At this point you have a good mental image of the expansion of the universe (raisin cake) and a good understanding of the cause of galaxy redshifts (stretched light waves). Next, we need to tackle the delicate topic of cosmic distances.

I said cosmologists use galaxy redshifts as distance indicators. Fine. But what do we actually mean by a galaxy's distance? Suppose a galaxy emitted a light wave long ago, when its distance from the Milky Way was 5 billion light-years. By the time the light finally arrives here on Earth, the distance may have grown to 10 billion light-years. After all, space has been expanding continuously.

Now here's a problem. The galaxy's redshift doesn't provide us with any information about the *original* distance. Nor does it provide us with information about the *current* distance. The only information we can glean from the redshift is how long the galaxy's light has been traveling through expanding space. That's neither 5 billion years nor 10 billion years. It's something in between, maybe 7 billion years or so.

So what can we say about the distance to the galaxy? Strictly, we should say something like, *This galaxy is so remote that its light, traveling through expanding space, took 7 billion years to reach us.* Quite a mouthful. So for pure simplicity, most astronomers would just say, *The galaxy is 7 billion light-years away.* After all, the 7-billion-year travel time for the galaxy's light is the only thing we can measure.

But of course, it's rather sloppy language. The next time someone talks to you about a galaxy at a distance of 11 billion light-years, remind yourself of what he or she means to say: the galaxy's light took 11 billion years to reach us—that's the only thing we can say for sure by measuring the galaxy's redshift. At the time the light was emitted (11 billion years ago), the galaxy was much closer to us—maybe just a few billion light-years. And right now, when the galaxy's light has finally arrived on Earth, that galaxy may be well over 20 billion light-years away.

Using the power of the Hubble Space Telescope, astronomers have succeeded in imaging thousands of remote galaxies whose light took many billions of years to reach us. This Hubble Ultra-Deep Field offers a view of the very early youth of the universe.

Now, I can almost hear you protest. Twenty billion light-years? How can a galaxy be so far away if the universe is only 13.8 billion years old? Didn't Einstein teach us that nothing can move faster than light? If so, how can anything have moved to a distance of 20 million light-years in less than 14 billion years' time?

But again, remember that the expanding universe is not a fireworks show. Galaxies are not racing through space. Instead, their distances grow because space itself is expanding. A very distant galaxy has never moved any faster than the speed of light, even though its distance from us may have been increasing by more than 300,000 kilometers every second. So there's no violation of Einstein's cosmic speed limit.

You may think this sounds absurd, but it's true. No energy, matter, or information of any kind is transported through space any faster than the speed of light. The General Relativity Highway Patrol has no tickets to issue. Still, the separation between two distant locations in the expanding universe can grow by more than 300,000 kilometers per second.

Does this mean that the universe expands faster than the speed of light? Yes and no. It all depends on the distance you're focusing on. Surprising as it may seem, the universe doesn't have a single expansion velocity. The distance between two relatively close points in space may grow at a rate of 10,000 kilometers per second. Meanwhile, the distance between two widely separated points in space may grow at a rate of 500,000 kilometers per second. And Einstein doesn't care.

In fact, for all astronomers know, the universe might well be infinite in size. That's hard to imagine—our brains aren't designed to cope with infinity. Then again, a finite universe is also hard to imagine, maybe even harder than an infinite one. If the universe were finite in size, wouldn't it need to have an edge? What would that edge look and feel like? What would lie beyond?

To ease your mind, let me briefly explain how the universe might *in principle* be finite without having an edge. If that sounds like a paradox, think about the two-dimensional model of a universe we discussed in Chapter 4—the sheet of graph paper. As three-dimensional beings, we can curve the paper into the surface of a sphere. Moving across this curved surface, imaginary Flatlanders who populate the two-dimensional universe never encounter an edge. Nevertheless, their world is finite in size—if they decided to paint it all yellow, they wouldn't need an infinite amount of paint.

Likewise, our three-dimensional universe could in principle be finite without having a well-defined edge if it's somehow curved in a higher dimension. If this is giving you a headache, don't worry—all available evidence suggests that our universe does *not* have any large-scale, global curvature. In that case, it might be truly infinite. (Which may give you another headache.)

But how can an infinite universe have grown from one single point? Well, it hasn't.

Here's the second big misconception: that all of the universe was concentrated in one single point at the time of the big bang. It was not, at least not if the universe is infinite in size, which I will assume from now on. A couple of billion years ago, the expansion of space hadn't yet reached its current level. All cosmic distances were half the size they are now. Galaxies were closer to each other. The average density of matter in the universe was eight times higher (if distances are half as much, volumes are one-eighth as much: $1/2 \times 1/2 \times 1/2$). But back then, the universe was also infinite in size. After all, as you may remember from high school, infinity divided by 2 is still infinity.

Much longer ago, around the time galaxies had just started to form, all cosmic distances were a tenth the size they are now, and the density of the universe was a thousand times higher (the volume was $1/1,000$ of what it is now: $1/10 \times 1/10 \times 1/10$). But infinity divided by 10 is also infinity. So again, the universe was infinite in size back then.

Almost 13.8 billion years ago, just a few hundred thousand years after the big bang, cosmic distances were about $1/1,000$ (a tenth of a percent) of their current values. Back then, galaxies or stars hadn't formed yet. The universe was filled with hot neutral gas, mainly hydrogen and helium. The universe's density was a billion times higher than it is now (because its volume was a billionth of what it is now: $1/1,000 \times 1/1,000 \times 1/1,000$); its temperature was a few thousand degrees Celsius. The whole universe was glowing as hot and brilliant as the surface of our Sun. But yes, even then it must already have been infinite in size.

Go back even further, and both the density and the temperature of the universe rise to extreme levels—so high, in fact, that there can't be any neutral atoms, not even protons or neutrons (see Chapter 5), just a seething broth of elementary particles and high-energy photons.

So what about the big bang?

Well, in a sense this *was* the big bang. When cosmologists talk about the big bang, they're usually referring to this superdense, superhot initial state of the universe. They can calculate what the universe looked like when it was ten years or one year or three minutes or a fraction of a second old. Pretty amazing. But just before they reach time zero, their theories break down. The true origin of the universe is still a mystery. For now.

Here's another way of looking at this. Long, long ago, densities and temperatures were extremely high *everywhere* in the universe. Each and every point of space once experienced the universe's superdense and superhot initial state. If we used a time machine to travel back 13.8 billion years to the same point in space that we occupy right now, we would be fried by the primordial plasma, as would be the case at every other location in the universe. By now, you probably guess what I'm getting at: *the big bang happened everywhere.*

So far, we've got galaxies pushed to larger separations like raisins in rising dough, light waves stretched during their billion-year trip through expanding space, a universe that may always have been infinite in size, and a big bang that happened everywhere. And I still have to introduce the afterglow of creation, which is important for our story on Einstein waves.

Before I do that, however, I need to tell you about the cosmic horizon. The cosmic horizon determines how far we can look out into our universe.

You might naively expect that astronomers can look as far as they want. Give them a bigger telescope, and they should be able to observe more distant galaxies, right? But that doesn't account for the finite speed of light—and the finite age of the universe.

What does the speed of light (300,000 kilometers per second) have to do with how far we can look into space? Everything. The reason: looking far into space also means looking far back in time.

On a clear summer night, you may have seen the bright star Deneb in the constellation Cygnus, the Swan. Deneb is a luminous giant star—it can easily be seen with the naked eye despite the fact that it's some 2,600 light-years away from us. That distance means that Deneb's light takes 2,600 years to reach Earth. In other words, the light we see tonight was emitted 2,600 years ago, around the birth of the Greek philosopher Thales of Miletus. We don't see Deneb as it is *now* but as it was more than two and a half millennia ago. We're looking back in time.

(In case you were wondering, yes, this means that Deneb might not exist anymore. If the star exploded in the year AD 400, the light from the explosion won't reach us for another thousand years or so.)

Now take the galaxy NGC 474, which we encountered earlier. Its distance is some 100 million light-years. The light we receive today was emitted when dinosaurs still roamed the surface of the Earth. Looking at NGC 474, astronomers are seeing a full 100 million years back in time. For more distant galaxies, this so-called lookback time can be billions of years. No wonder telescopes are sometimes called time machines!

The good thing about looking back in time is that cosmologists can study the evolution of the universe. Do you want to know what the universe looked like some 8 billion years ago? Just train your telescope at galaxies 8 billion light-years away (or, more precisely, at galaxies that are so remote that their light, traveling through expanding space, took 8 billion years to reach us.) Ten billion years ago? Just look out a little farther.

The downside is that there's a fundamental limit to how far out we can look into space. If our universe was born 13.8 billion years ago, that's also the maximum amount of time a light ray can have been traveling. So we can't look back any further than 13.8 billion light-years—it's as simple as that. The universe may be infinite, but we can observe only a relatively small part of it: a spherical region with a radius of 13.8 billion light-years, centered on our own Milky Way. That's what we call the observable universe. The surface of the sphere is our cosmic horizon.

Several things are worth mentioning here. One: You may have noticed that I've chosen to stick to the sloppy convention of converting light travel times into distances. In fact, the actual *current* radius of our cosmic horizon is some 42 billion light-years. But it's really very convenient to use the one-on-one relation between distance and look-back time.

Two (very important): The cosmic horizon is a *fundamental* limit to our observational capability. No telescope, regardless of its size and power, will ever reveal what's out there at larger distances. It's just not possible, period.

Three: As the universe ages, the *observable* universe grows in size. Every year, its radius grows by one additional light-year. Unfortunately, in the end, the growth of the observable universe won't be able to keep up with the expansion of space, which is actually accelerating (more on that in Chapter 16).

Four: Every location in the universe has its own cosmic horizon, of course. Think of ships on an ocean. Every ship has its own ship-centered horizon, beyond which the ship's sailors can't see. Likewise, every observer in the universe is in the center of his or her own small, personal *observable* universe.

Five: Our cosmic horizon is not a physical entity. An alien observer located exactly at our horizon would see nothing peculiar there. His immediate surroundings would look like ours, with aging stars and mature galaxies. After all, he lives in the same 13.8-billion-year-old universe as we do. But we would be on *his* horizon, too. Looking across the vastness of space in our direction, our alien friend would look 13.8 billion years back in time, to an epoch well before the birth of our Milky Way galaxy, let alone the Sun and Earth.

Now we're finally getting to the afterglow-of-creation part—a term coined (or at least popularized) by British astronomy writer Marcus Chown. The afterglow of creation is what an observer sees at the very edge of his or her personal observable universe: a fading image of the violence of the big bang.

The farther you look into space, the further back you look in time. At the cosmic horizon—at the edge of the observable universe—the lookback time is 13.8 billion years. Any radiation we receive from that remote edge was emitted 13.8 billion years ago, just after the birth of the universe. It provides us with a view of what the universe looked like back then.

For the first few hundred thousand years of the existence of the universe, it was filled with a seething plasma so dense and hot that light could not propagate through it. But when the universe was 380,000 years old, the density and the temperature had dropped enough for neutral atoms to form. For the first time, photons (Einstein's "particles of light") could travel unhindered through space. The universe had become transparent.

Back then, as I said before, the whole universe was glowing as hot and brilliant as the surface of our Sun. So if we look back to that time, as we do at the cosmic horizon, we should observe this big bang glow in every direction we choose to look. And we do! But this primordial radiation has traveled for 13.8 billion years (minus a negligible 380,000), and it has become exceedingly faint. Moreover, the radiation has traveled through an ever-expanding universe. As a result, the wavelength got stretched by a factor of about 1,000. So instead of a dazzling glow at visible wavelengths, what's left is an almost imperceptible hiss at radio wavelengths. The hiss is commonly known as the cosmic microwave background, or CMB for short. But if I'm in an eloquent mood, I prefer to call it the afterglow of creation. Sounds more poetic.

The cosmic microwave background was discovered in 1964, more than half a century ago. Since then, it has been studied in ever-increasing detail, as we'll see in Chapter 10. This is not surprising, of course: the CMB is the oldest signal astronomers can observe. It's the closest we can get to witnessing the birth of the universe.

Some people find it confusing that we can study the afterglow of creation for decades on end. In fact, if instruments had been sensitive enough back then, the Neanderthals or even the dinosaurs could have observed the CMB. Likewise, our distant descendants may still study it millions of years from now. But wasn't the birth of the universe a fleetingly brief event? Wouldn't you expect the radiation from that epoch to just zip by? How is it possible that we continue to see the afterglow?

This also has to do with the finite speed of light. To understand this, let's do another thought experiment. Imagine we're on a large city square with many thousands of other people. It's really crowded. We've all been told to synchronize our watches to the nearest second and to shout "Boo!" at exactly noon. Oh, and one other detail: the city square is on a planet where the speed of sound is not the regular 330 meters per second but just 1 meter per second.

So what happens at noon? You shout "Boo!" as loud as you can. The sound you produce is spreading out in all directions. Within a second, you can't hear your own shout anymore. But at 12:00:01, you hear the collective "Boo!" from the people around you at a distance of 1 meter—the sound *they* produced at noon took one second to reach you. And at 12:00:02, you still hear "Boo!"—from people at a distance of 2 meters. Even a full minute after noon, shouting will still reach your ear, from people 60 meters away from you.

The funny thing is, no one is shouting anymore. Everyone on the city square just produced one, brief "Boo!" at exactly noon. But you keep hearing shouts coming in from ever-larger distances. If the square were really big, you could hear the noon "Boo!" for hours on end. And the same would be true for every other person on the square. After all, someone 300 meters away from you will hear *your* "Boo!" at 12:05. And so forth.

The city square is the universe. The collective "Boo!" at noon is the relatively brief burst of cosmic microwave background emitted shortly after the big bang. The afterglow radiation that was produced 13.8 billion years ago at *our* point of space, right here under our feet, has long since dispersed throughout the universe. But we're still receiving the faint signal from ever more distant points in space. (If you want to improve the analogy even more, replace the city square pavement with a rubber sheet and have someone pull at the edges— there's your expanding universe!)

Cosmology is a vibrant field of science, full of unsolved mysteries and exciting discoveries. Maybe we will never really understand how everything began, but we're well beyond Terry Pratchett's "In the beginning there was nothing, which exploded." And who knows, the detection of primordial gravitational waves, produced at the time of the big bang, may open up a new window on the birth of the universe. So now join me on a trip to the geographical South Pole to learn how close we are to that long-awaited breakthrough.

(((10)))

Cold Case

I'm meeting Shaul Hanany in the Antarctica Hilton.

No, this is not a fancy hotel with a lobby, a bar, and valet parking for snowmobiles. "Antarctica Hilton" is just a nickname for what's basically a shed to temporarily keep people from freezing to death. One door, a couple of windows, two wooden benches—that's all. It's surrounded by vast stretches of windblown ice and snow in every direction. And no heating, by the way.

Earlier that day, I had visited the National Science Foundation's Long Duration Balloon Facility (LDBF), not too far from McMurdo Station, the US research base on the coast of the frozen continent. In a huge hangar on giant skis—officially called a Payload Assembly Hall—the BLAST telescope was being readied for its fifth and final mission. BLAST, led by Mark Devlin of the University of Pennsylvania, is short for Balloon-Borne Large-Aperture Submillimeter Telescope. As the name implies, it studies the universe at wavelengths just shy of 1 millimeter. That's not possible from the ground (the microwave radiation is absorbed by water molecules in the atmosphere), so Devlin and his team use a giant helium balloon to take the instrument up into the stratosphere for about two weeks on end.

In a second Payload Assembly Hall, Hanany had been checking on the progress of his own balloon experiment called EBEX (which simply stands for "E and B experiment"). At the end of the afternoon, LDBF camp manager Scott Battaion drives both of us back in his heavy-duty pickup truck. But not all the way to McMurdo—that would cost him too much time. He drops us off at the junction with the ice road to Pegasus Field, the base's main airstrip. In the Antarctica Hilton, we can await the arrival of Ivan the Terra Bus, the monstrous McMurdo airport shuttle with its oversized tires.

Shaul Hanany is a physicist at the University of Minnesota. Waiting at the Hilton together for almost an hour gives me ample opportunity to ask him everything about EBEX. Meanwhile, Hanany becomes ever more nervous. What if the shuttle never shows up? We're all alone on the ice, without any means of communication.

EBEX is designed to measure the polarization of the cosmic microwave background. Today's preflight tests looked promising, Hanany tells me. Launch is scheduled within two weeks or so. During its mission, EBEX *might* make a revolutionary discovery. Hanany and his colleagues hope to find, hidden in the polarization patterns of the big bang radiation, the fingerprints of primordial gravitational waves dating back to the very origin of the universe.

Ivan the Terra Bus eventually shows up as a red dot on the all-white horizon. The driver—everyone calls him Shuttle Bob—tells us he got stuck in a snowbank, hence the delay. Half an hour later we're back at McMurdo Station, which has the look and feel of a military base.

I spent a week at McMurdo in December 2012 as one of three selectees in that season's Antarctic Journalist Program of the National Science Foundation. It was a great experience. I met geologists, penguin researchers, climatologists, meteorite hunters (including NASA astronaut Stan Love), glaciologists, astrophysicists—you name it. I visited Scott's Hut, erected in 1911 by English explorer Robert Falcon Scott before he and his men set off on their fatal expedition to the South Pole. I climbed 230-meter-high Observatory Hill, which carries a memorial cross for the deceased explorers. I visited the local chapel and talked to Michael Smith, the McMurdo priest. I listened to a tipsy astrobiologist singing off-key karaoke at Gallagher's Bar (no,

I won't disclose any names). Our small group made a helicopter trip to Cape Royds and Cape Evans, we visited the test site of the Wissard ice-drilling project, and we joined a hike across photogenic ice pressure ridges. And yes, I made my way to the Long Duration Balloon Facility.

But by far the most spectacular part of the visit was the day trip to the Amundsen-Scott South Pole Station on December 10. While McMurdo is on the coast of the Antarctic continent, close to the Ross Ice Shelf, Amundsen-Scott is at the geographical South Pole—the most southerly point on the globe. It's a three-hour flight in the *Spirit of Freedom*—a propeller-driven, ski-equipped Lockheed LC-130 military airplane. It's a balmy day at the pole, with a wind chill of just minus 37 degrees Celsius. Still, despite my ECW gear (ECW is military shorthand for "extreme cold weather"), I don't dare to walk the kilometer or so from the station's living quarters to the Dark Sector, where most astronomy-related experiments are concentrated. We're taking the tracked snowmobile instead.

One of the most impressive buildings at the Dark Sector is the IceCube Laboratory. Even more impressive is the fact that this structure is dwarfed by the invisible scientific experiment it supports. IceCube is the world's largest neutrino observatory, but you can't see it. It consists of more than 5,000 ultrasensitive light detectors frozen into a cubic kilometer of subsurface Antarctic ice. The lab building houses only its powerful computer systems. IceCube's photo detectors register extremely rare flashes of light in the dark, transparent ice caused by passing neutrinos from the cosmos. (I introduced neutrinos in Chapter 5. They're elusive subatomic particles that were produced in large numbers during the big bang. They also play an important role in supernova explosions.)

Not far from the IceCube Laboratory is the equally impressive Martin A. Pomerantz Observatory. It's named after a pioneer in Antarctic astronomy who died in 2008. At one end of the elongated, two-story building is the 10-meter dish of the South Pole Telescope; at the other end is the conical collar that houses the BICEP2 instrument (BICEP stands for Background Imaging of Cosmic Extragalactic

The BICEP2 telescope is located in the cone-shaped "collar" on the roof of the Martin A. Pomerantz Observatory, very close to the National Science Foundation's Amundsen-Scott South Pole Station at the geographical South Pole. At the left is the 10-meter South Pole Telescope.

Polarization). Both study the cosmic microwave background—the faint cosmic radio waves that date back to 380,000 years after the big bang. The collar prevents stray radiation from human activity from entering the sensitive telescope.

With its 26-centimeter lens, BICEP2 is smaller than quite a few amateur telescopes. But its focal plane is cooled to just a quarter of a degree above absolute zero and filled with 512 extremely sensitive superconducting sensors to register every single microwave photon from the sky. Like Shaul Hanany's EBEX instrument, BICEP2 studies the polarization of the cosmic microwave background. (At least it did in December 2012. It has since been replaced by an even more powerful instrument.)

————

The cosmic microwave background (CMB) was accidentally discovered in 1964 by American radio engineers Arno Penzias and Robert Wilson. It's the universe's oldest "light." Astronomers cannot look further back in time than to the CMB epoch, because the universe was too hot, dense, and opaque in the first 380,000 years of its existence for electromagnetic radiation to freely propagate through space. So this is the closest we can get to the big bang.

As I described in Chapter 9, what started out as a blinding bath of high-energy radiation has by now faded and cooled down to an almost imperceptible hiss at millimeter wavelengths, corresponding to a temperature of a mere 2.7 degrees above absolute zero. If you want to understand the origin of the universe, this cold echo is what you need to study in detail, no matter how hard that might be.

It's difficult not just because the cosmic microwave background is so faint, but also because microwaves from space are readily absorbed by water molecules in the Earth's atmosphere. (The same process ensures that water-rich food heats up quickly in a microwave oven: the water molecules absorb almost all the radiation's energy.) So by far the best place to observe the CMB is space itself. Indeed, the best all-sky maps of the CMB have been produced by three successive space missions.

The first CMB space mission—and arguably the most revolutionary one—was COBE (Cosmic Background Explorer). This was the mission that grew out of Rai Weiss's early work at MIT. Launched in November 1989, COBE was the first to reveal small temperature differences in the background radiation at a level of a ten-thousandth of a degree. The tiny "hot" and "cold" spots correspond to regions of slightly higher and slightly lower density in the very early universe, which eventually gave rise to the origin of galaxies and clusters. Without those primordial density fluctuations, the current universe would be a dark and boring ocean of hydrogen and helium with a density of about one atomic nucleus per cubic meter or so. There would be no galaxies, let alone stars, planets, or people. We owe our existence to those small perturbations. COBE principal investigators John Mather and George Smoot received the 2006 Nobel Prize in Physics for their breakthrough findings.

Much more detailed maps of the CMB were later produced by NASA's WMAP satellite (Wilkinson Microwave Anisotropy Probe, launched in June 2001) and by the European Space Agency's Planck mission, named after famous German physicist Max Planck. The Planck mission was launched in May 2009 and was operational until October 2013. Both WMAP and Planck have revealed a wealth of information on the early universe. In a sense, they turned cosmology into a precision science.

The CMB can also be observed from the ground—not at sea level, of course, because of the absorbing effect of Earth's atmosphere, but from any place that's high and dry enough. Take your microwave telescope to a site where you leave the bulk of the atmosphere's water vapor below you, and you're in business.

The South Pole is one of those unique locations. The Amundsen-Scott South Pole Station is actually at an altitude of 2,835 meters above sea level. Moreover, the cold Antarctic sky is extremely dry (technically, Antarctica qualifies as a desert), so there's little atmospheric water vapor. In 1999, University of Chicago scientists chose this location to build their Degree Angular Scale Interferometer (DASI)—an eye-catching instrument with thirteen individual detectors on a single mount. BICEP1—a small precursor of BICEP2—started operations in 2006. Construction of the 10-meter South Pole Telescope was completed in early 2007.

Another great location is the Llano de Chajnantor in northern Chile. This high-altitude plateau, over 5,000 meters above sea level and surrounded by volcanoes, is now home to the sixty-six-dish Atacama Large Millimeter/submillimeter Array (ALMA). Going up there is a literally breathtaking experience. In November 2004, during my third visit to Chajnantor, work on ALMA hadn't started yet. Even the road to the observatory site was still under construction. But back then, the DASI-like Cosmic Background Imager was already in operation. Three years later, construction of the 6.5-meter Atacama Cosmology Telescope was almost finished. Like BICEP2, the actual instrument is surrounded by a huge conical shield to keep out stray radiation, but if you hike to the 5,600-meter summit of nearby Cerro Toco, which I did in 2013, you have a great view of the instrument.

———

Many space missions and ground-based instruments are keeping an eye on the cosmic background radiation—the afterglow of creation or, as it is sometimes called, the baby photo of the universe—but the latest experiments all focus on the CMB's polarization. One of the many holy grails in cosmology is detecting the elusive B-mode patterns in the polarization of the CMB, produced by primordial gravitational waves that date back to the inflationary era of the very early universe. That's a lot of jargon, but I'll walk you through it step by step.

Let's start with polarization. Light is an electromagnetic wave phenomenon, as James Clerk Maxwell discovered in the late nineteenth century. Normally, the fluctuating electric and magnetic fields oscillate equally strongly in every direction: horizontal, vertical, diagonal, and everything in between. But as soon as a light wave is reflected, it becomes polarized: the oscillations are stronger in one direction than in the other.

Polarized sunglasses make smart use of this effect. When sunlight reflects off a flat surface, such as water, snow, or a road, it becomes horizontally polarized to a certain degree: the reflected waves oscillate much more strongly in the horizontal direction than in the vertical direction. Polarized sunglasses preferentially block the horizontal oscillations, so the reflections become much dimmer. If you look through your polarized sunglasses with just one eye, the effect is clearly visible when you rotate the glasses by ninety degrees.

Photographers also know about polarization. Sunlight is scattered by air molecules and dust particles. It becomes slightly polarized as a result. Putting a rotatable polarizing filter in front of your camera lens enables you to make the sky much darker, which usually makes for more dramatic pictures.

Of course, instead of filtering polarized light, you can also *study* its polarization to learn about the source of the effect. In the case of air pollution, for example, atmospheric physicists can measure the strength and the direction of the polarization of sunlight at various wavelengths. That tells them about the size, structure, and composition of the pollutant particles.

The cosmic microwave background has been traveling through the universe for 13.8 billion years. Since the space between the galaxies is an almost perfect vacuum, you wouldn't expect the CMB to be strongly polarized. Still, there is a very, very tiny effect. It turns out that the CMB is polarized by about 1/300,000 of a percent. This means that at every point in the sky, there's an incredibly tiny tendency for the microwave radiation to oscillate in a certain direction.

Such a small amount of polarization is hard to measure. Imagine sprinkling 60 million rice grains on the floor at random and then finding a slight preference in their orientation: 29,999,999 grains are oriented within 45 degrees of the east-west direction, while 30,000,001 grains are within 45 degrees of north-south. That's about the detection sensitivity you would need to measure the polarization of the CMB. This was first achieved in 2002 by DASI.

So how does this slight polarization come about? It's not because the background radiation has been reflected off stars or planets or has been scattered by interstellar dust particles. No, the tiny amount of polarization was imprinted on the CMB when it started its journey toward us, some 13.8 billion years ago. It's the "fingerprint" of the uneven distribution of matter in the very early universe. I already mentioned those tiny density fluctuations in the primordial gas—the "seeds" of the universe's current large-scale structure. They led not only to the small temperature variations in the CMB (the "hot" and "cold" spots first observed by COBE), but also to a very small amount of polarization in directions that vary across the sky.

So that's the polarization-of-the-CMB part. Now what about inflation, primordial gravitational waves, and B-mode patterns?

Inflation is what happened to the universe in the very first tiny fraction of a second of its existence. Or, I should say, it's what cosmologists *think* happened. It's far from a proven concept, or even a mature theory; rather, inflation is the common denominator for a bunch of hypothetical scenarios, one of which might be true. Or, as most cosmologists would say, one of which *has to* be true. That's because inflation is the only known way to solve a number of nagging problems with the original big bang theory.

I won't go into all the details here, but it all boils down to a very brief period of exponential expansion. Before the universe was 10^{-32} seconds old (that's 0.000,000,000,000,000,000,000,000,000,000, 001 seconds), space was blown up ("inflated") by a factor of 2 about 200 times in succession. As a result, every single distance between two separate points in space grew to some 10^{60} times its original value. At the end of this extremely brief inflationary epoch, the more familiar "linear" expansion of the universe took over at a much more sedate rate. To a certain extent, inflation can be compared to the very first stages of the growth of a fertilized human egg cell. At first, the number of cells grows as 1, 2, 4, 8, 16, and so on. But fortunately, this exponential growth stops after a while, turning into a much slower rate (otherwise you would now be larger in size than the observable universe).

There are good quantum mechanical reasons to believe in the "bang of the big bang," as inflation has been called. Moreover (and again without going into the details), it's the only imaginable way to explain why the observable universe looks so homogeneous and why spacetime doesn't seem to possess any global, overall curvature. The concept was first proposed in 1980 by Alan Guth, a theoretical physicist who was at Princeton University at the time. It has been extended and modified since, especially by Russian American physicist Andrei Linde at Stanford University. Inflation may be hard to grasp and even harder to believe, but most cosmologists have become very much used to it.

The various inflationary scenarios that float around differ only in the details: what exactly caused inflation, when it started, how fast the exponential expansion was, how long it lasted, how it stopped, et cetera. The problem, of course, is that we can't look back to a mere 10^{-32}

second after the birth of the universe to see what *really* happened. The cosmic microwave background is the oldest light from the early days of the universe that we can scrutinize, and it was produced 380,000 years after the event. So how can cosmologists ever hope to prove that inflation really happened, let alone discriminate between the various versions?

That's where gravitational waves come in. Inflation blows up *everything* in space. Subatomic quantum fluctuations in the newborn universe were expanded all the way into the density variations that left their mark in the cosmic microwave background. Likewise, quantum fluctuations in the gravitational field must have been blown up, too—not into more density fluctuations but into primordial Einstein waves that reverberated through the very fabric of spacetime. At least that's the theory. The amplitude of these primordial waves depends on the precise specifics of inflation.

So if we could detect primordial gravitational waves, we would have a strong indication that inflation really did happen. We might even be able to refute at least a few inflationary scenarios. Unfortunately, the inflationary waves can never be detected directly. After 13.8 billion years of cosmic expansion, they have wavelengths of hundreds of millions of light-years, and there's no way for us to measure them. But they, too, left their mark on the cosmic microwave background. Just as the CMB became slightly polarized by the density fluctuations in the early universe, it also became slightly polarized by interaction with the primordial gravitational waves.

Measuring the CMB polarization that is caused by inflationary Einstein waves could tell us about events during the very first split second after the birth of the universe. It would provide us with a unique way to look beyond the 380,000-year limit, right to the very beginning of space, time, matter, and energy. There's just one remaining problem: the polarization due to primordial gravitational waves is a thousand times smaller than the polarization due to density fluctuations, which is itself tiny. How can we ever hope to disentangle the two effects?

That's where the B-mode patterns come in, finally. Suppose an American pastry chef has provided you with many thousands of identical little cakes, each with a topping of whipped cream. You suspect there may be a handful of European cakes among them. But that's hard to find out: on both continents, they follow the same recipe, so the cakes all look the same. Then you learn—and I'm completely making this up—that European cooks, when they force cream out of a piping bag, always turn the bag around, either clockwise or counterclockwise. American cooks keep their piping bags still. So while the American toppings are perfectly symmetric, the European ones are a bit swirled, one way or the other. Now it's easy to identify them, even though the cakes themselves are otherwise identical.

Something similar is the case with the two types of polarization. If you create a map that shows the strength and direction of polarization for every point on the sky, you see certain patterns. For the polarization caused by density fluctuations, those patterns are symmetric—they don't possess a particular "handedness." These are called E-mode patterns. For the (much smaller) polarization caused by primordial gravitational waves, the resulting patterns on the sky display a small additional swirl, either to the left or to the right. These are called B-mode patterns. (The use of those letters goes back to Maxwell, who used E to denote electrical fields and B to denote magnetic fields.)

There's one minor complication: subtle B-mode patterns can also form when the polarized background radiation passes close to a massive cluster of galaxies. The gravitational lensing effect of the cluster—comparable to the relativistic bending of starlight by the Sun—leaves a small swirl in the otherwise symmetric E-mode patterns that has nothing to do with inflation or primordial gravitational waves. Fortunately, the B-mode patterns that are caused by gravitational lensing occur only on small scales of much less than a degree of the sky. They were first detected by the South Pole Telescope in 2013. So if you want to prove inflation and find evidence for the existence of Einstein waves from the birth of the universe, you have to search for much larger B-mode patterns at angular scales of at least a degree of the sky.

So now you know why Shaul Hanany's balloon experiment was called EBEX. Its goal was to disentangle the E- and B-mode patterns in the polarization of the cosmic microwave background. The discovery of large-scale B-mode patterns would imply the existence of primordial gravitational waves and thus confirm the theory of cosmic inflation. The relative "strength" of the B-mode patterns would provide some information on the precise cause and timing of inflation.

EBEX was launched on December 29, 2012. Its balloon-borne mission lasted for about two weeks. It collected a lot of interesting data, but it did not discover B-mode patterns. But during my Antarctica visit, the BICEP2 telescope, behind its conical shield, was busily collecting polarization measurements of a large swath of southern sky. As more and more data were collected over the months, the patterns in the CMB polarization became more and more evident. Eventually, in the course of 2013, the BICEP2 team started to get excited— cautiously at first, but with more and more confidence over the months. It looked like they had finally found the elusive large-scale B-mode patterns: the long-awaited "proof" of inflation and, more important to our story, the first clear fingerprint of gravitational waves dating back to the first split second of cosmic history.

On Wednesday, March 12, 2014, the Harvard-Smithsonian Center for Astrophysics (CfA) in Cambridge, Massachusetts, issued a brief statement: on Monday, March 17 at noon it would host a press conference "to announce a major discovery"; no further information was given. CfA's cramped Phillips Auditorium could hold only a limited number of reporters, but public affairs officers David Aguilar and Christine Pulliam arranged for a live video stream. According to their IT team, the Harvard servers could easily handle a thousand simultaneous viewers, so everything appeared to be good to go.

What Aguilar and Pulliam did not realize is that the rumors that started to fly about the announcement on social media (Big bang! Inflation! Gravitational waves!) created an enormous amount of buzz, and so many people logged on that the webcast crashed right at the

The BICEP2 team presents its results at a press conference on March 17, 2014, at the Harvard-Smithsonian Center for Astrophysics in Cambridge, Massachusetts. *From right to left:* John Kovac, Chao-Lin Kuo, Jamie Bock, and Clem Pryke. At the far left is independent commentator Marc Kamionkowski.

start of the press conference. It took quite a while before a backup remote viewing option was available, but that, too, had insufficient bandwidth.

CfA's John Kovac, the principal investigator for BICEP2, recalls that he had never experienced something like this before. I assume he also had never been the first presenter at such a major press conference. If he had, he would have started with the main message right away. Instead, he gave a mini-lecture on the history of CMB observations. Behind him, a title slide carried the relatively obscure phrase "Detection of B-mode Polarization at Degree Scales Using BICEP2." Only webcast viewers with a lot of background in cosmology would understand what this was all about.

Kovac's three project coleaders only made matters worse. Chao-Lin Kuo from Stanford University tried to explain inflation and how it would produce primordial gravitational waves and B-mode patterns in the polarization of the cosmic microwave background. "It's a pretty hard concept to explain," he said. (I fully agree.) Jamie Bock from Caltech and NASA's Jet Propulsion Laboratory in Pasadena gave a rather incomprehensible talk about detector technology. And Clem Pryke, one of Shaul Hanany's colleagues at the University of Minnesota, discussed the data analysis. All in all, you wouldn't get the impression that a revolution in cosmology was unfolding.

But then the mood changed. Aguilar and Pulliam had invited theoretical physicist Marc Kamionkowski from Johns Hopkins University in Baltimore, Maryland, to comment on the results being presented. Kamionkowski was the only person behind the table *not* wearing the black BICEP2 T-shirt to underscore the fact that he was an impartial outsider. The first sentence of his prepared statement ended up in many newspaper stories the next day.

"It's not every day," he said, "that you wake up and learn something completely new about what happened one trillionth of a trillionth of a trillionth of a second after the big bang." Kamionkowski described BICEP2's discovery as "really cool" and as "cosmology's missing link." And he went on: "This is not just a home run; it's a grand slam. It's a smoking gun for inflation . . . and it's also the first detection of gravitational waves. . . . If the results hold up, inflation has sent us a telegram, encoded in gravitational waves and transcribed on the microwave background sky."

Now everyone was awake. Alan Guth and Andrei Linde—the two main originators of inflation theory—were in the auditorium, and they happily expanded on the mind-boggling fact that almost every inflationary scenario implies the existence of parallel universes. As Linde said to reporters, "Evidence for inflation will be pushing us in the direction of taking the multiverse seriously."

"Space Ripples Reveal Big Bang's Smoking Gun," wrote the *New York Times* on its website that same day. *National Geographic*'s headline read, "Big Bang Discovery Opens Doors to the 'Multiverse.'" BBC News quoted Alan Guth as saying that the experiment is Nobel Prize worthy. To the British weekly magazine *New Scientist,* Harvard University theoretician Avi Loeb described the result as "the most important breakthrough in cosmology over the past 15 years." Christine Pulliam collected some 3,500 news clippings about the announcement. The BICEP2 website hosting the team's scientific publications received more than 5 million hits within a couple of days.

A short YouTube video of Chao-Lin Kuo bringing the news to his Stanford mentor, Andrei Linde (filmed well before the press conference), went viral. Linde and his wife, Renata Kallosh (herself also a theoretical physicist), are visibly moved by Kuo's message. As they open the door of their house, Kuo just says, "So I have a surprise for you. Five sigma, it's 0.2," referring to an unexpectedly strong signal at a high level of statistical significance. Kallosh embraces him; shortly thereafter, champagne flows.

Still, the hype and excitement had a major downside, too. The many caveats surrounding the BICEP2 announcement often got lost, at least for the general public. This was certainly not the fault of the scientists. Every single researcher at the meeting and every expert interviewed by journalists had the same message: "*if* this is really true," "*if* the results hold up," "*if* confirmed by other experiments," "this requires further scrutiny." But a lot of people skipped over the cautions and focused on "big bang," "breakthrough," "multiverse," and "Nobel Prize."

And "gravitational waves," of course. As Kamionkowski had made very clear, this would be an important first, achieved just one year before the centenary of Albert Einstein's theory of general relativity. True, the waves hadn't been detected directly, but the indirect evidence for their existence was almost as convincing as it had been in the case of the famous Hulse-Taylor pulsar and other binary neutron stars. Back in the 1970s, scientists had inferred the existence of gravitational waves much as they would infer the existence of a thief from the fact that stuff was missing and the door of the house stood open. Now, though, they had found the metaphorical footprints of the perpetrator in the flowerbed.

If confirmed.

Right after the press conference, other theoretical physicists aired their concerns. The reported B-mode signal was much stronger than anyone had expected. The implications didn't seem to agree too well with preliminary results from other experiments. And could the

BICEP2 team really be sure that the observations had only one possible explanation?

Those were valid concerns, as John Kovac and his colleagues knew only too well. BICEP2 had been studying a region of the sky well away from the plane of our own Milky Way to minimize the risk of foreground contamination. The reason: dust particles in the Milky Way also emit microwaves, and in the presence of magnetic fields, those waves can exhibit a small amount of polarization, with B-mode patterns and all. If measurements had been available at a number of different wavelengths, it would have been easier to correct for this potential problem. Unfortunately, the BICEP2 detectors were sensitive only to one particular wavelength: 2 millimeters, corresponding to a frequency of 150 gigahertz.

In order for the team to convince themselves that they had not been tricked, they used the best available information on the distribution of Milky Way dust that they could get their hands on. They would also have preferred to check their results against the newest sensitive dust measurements of ESA's Planck satellite. Indeed, Kovac had approached the Planck team to suggest a joint analysis of the two datasets. But the Planck scientists politely told him to wait until they publicly released their observations, which might well take another year or two.

Meanwhile, the BICEP2 telescope had been dismantled in early 2013. The experiment was completed; the data analysis was almost done. Should they wait another couple of years before presenting their results to the world, with the risk of being scooped by others? Or would it be best to tell colleagues what they had found?

The answer to the dilemma presented itself in April 2013, during a conference at the European Space Research and Technology Centre in Noordwijk, the Netherlands. The conference was called "The Universe as Seen by Planck," and it provided an in-depth look at the mission's initial scientific results. On the second day of the conference, the Planck team showed their preliminary maps of the distribution of galactic dust and of its polarized emission.

A map is a visual representation of quantitative science data—it's not the real thing. The results were preliminary, too. And a smartphone photo made of a projected PowerPoint slide is not the best material to work with. But it's better than nothing. The BICEP2 team decided to move ahead. Eventually they prepared a paper for *Physical Review Letters*. And in early January 2014, Kovac approached his institute's public affairs office. Might the news warrant a press conference?

Normally, a university or research institute doesn't go public with a scientific result before the corresponding paper has been accepted for publication. That will happen only after one or more anonymous reviewers have had a chance to carefully read it and give their professional comments. (You may remember that Albert Einstein was appalled by this peer review process back in the 1930s.) But the public affairs officers of the Harvard-Smithsonian Center for Astrophysics decided not to wait that long. They were concerned that the news would leak. Instead, they organized a brief BICEP2 scientific symposium on the morning of March 17, 2014, in the same Phillips Auditorium, and tied the press conference to that meeting.

Alas, the results did not stand the test of time, as the early critics had expected was likely. As soon as other scientists dove into the results—which had been published on the project's website on the day of the press conference—they encountered serious issues with the way Kovac's team had handled dust contamination. It soon became clear that some of the bolder BICEP2 claims had to be taken with a grain of salt (or, perhaps, dust). Later that year, the joint analysis with the Planck team finally got started, and the original BICEP2 paper had to be partly rewritten. The bottom line of this analysis was a much smaller value for the relative strength of the large-scale B-mode patterns. Given the uncertainty in the measurements, you can't even exclude the possibility that they're not there. Ergo, no convincing evidence for the existence of primordial gravitational waves. No proof for inflation. No revolution.

Not yet, in any case.

———

Looking back at the episode, John Kovac doesn't regret the chain of events too much. Science is a never-ending process of collecting more data and adapting your conclusions, he says. Moreover, the researchers themselves had always been very clear about the uncertainties and the possible pitfalls, so their professional reputation had not suffered. "But," he adds, "we've learned some important lessons about science communication in the Internet age. You really have to be very, very clear on everything you do and say. You also need to make sure that all potential caveats are clearly communicated." (And, I might add, that your webcast has enough bandwidth.)

At the time of writing, it's almost three years since the BICEP2 results were first announced. In the meantime, the Keck Array, consisting of five BICEP2-like telescopes on a single mount, has been scanning the skies at two frequencies for a number of years. In fact, it was already in operation when I visited Antarctica in December 2012. And since May 2016, a larger and more efficient instrument, called BICEP3, has joined the search from the southern tip of the Earth. BICEP3 has a 68-centimeter aperture and contains no fewer than 2,560 individual microwave detectors.

Next door, the much-larger South Pole Telescope is now equipped with a polarization-sensitive camera. The same is true for the Atacama Cosmology Telescope at Llano de Chajnantor in northern Chile. Many smaller instruments are operational, too, with funny names like QUIJOTE, POLARBEAR, AMiBA, and CLASS. Chinese astronomers are building a new microwave polarization telescope in Tibet. And, of course, there are new balloon-borne experiments to follow up on Shaul Hanany's EBEX experiment, such as Spider and PIPER. Anytime now, one of those projects might claim the first detection of large-scale B-mode patterns—and, consequently, of gravitational waves produced at the very birth of the universe.

The race has become more competitive than ever, says Kovac. At the same time, he adds, it has also become more *cooperative* than ever. Many scientists are involved with more than one experiment. Various teams are working together in analyzing their data. And the community as a whole is now charting plans for the future. Who knows—a couple of years from now, it may be time to start thinking about a new space mission.

The BICEP2 episode was very instructive for the scientists involved. But the LIGO and Virgo teams learned something, too. Ever since the start of gravitational wave experiments, back in the 1960s, there had been uncertainties, false claims, and retractions, all of which had caused great embarrassment. The original hype surrounding the premature announcement of the BICEP2 results had done nothing to improve the shaky reputation of the field. The LIGO and Virgo teams decided that they would not announce the detection of gravitational waves from space unless they were absolutely certain about their claims and the results had been peer-reviewed. Even then, communication with the media and the public at large would have to be professionally managed and precisely controlled.

Advanced LIGO was almost ready to start operating. Both the Hanford and the Livingston detectors had reached "full lock"—the interferometer equivalent of "first light" for an optical telescope. The initial commissioning was done. Scientists, engineers, and technicians were carrying out final tests and checkups. The two detectors had been brought online in engineering mode, as it was called. Friday, September 18, 2015, would see the official start of Observing Run 1.

Meanwhile, LIGO scientists were polishing the protocols: what to do if a detection was made, how to check on its authenticity, when to alert the press, why it was important to say nothing to anyone until there was absolute certainty about any claim. There were rules and guidelines for everything. After all, given the enhanced sensitivity of the advanced detectors, the first Einstein waves could conceivably be picked up within weeks or months.

Possibly.

Hopefully.

Gotcha

A ripple in the fabric of spacetime is racing through the universe. It's a tiny disturbance in the four-dimensional metric. It changes the local curvature ever so slightly this way and that. Over the past 1.3 billion years, it has weakened tremendously. But it's still there, a faint reverberating echo of a dramatic event, just as a clap of thunder slowly dies away in the distance long after the lightning has struck.

The gravitational wave is not alone. Many similar undulations propagate through the universe—in every possible direction and with a wide range of frequencies and amplitudes. They have done so for billions of years. Almost imperceptibly, spacetime is constantly quivering like a drumhead. But this particular wave is a special one. It is destined to be the first one in the history of the universe that will actually be detected by humans.

Speeding through space at 300,000 kilometers per second, the wave entered our galaxy about 100,000 years ago. It sent tiny shivers through stars and planets as it moved through the Milky Way in our direction. In 1915, the year in which Albert Einstein formulated his theory of general relativity, it had only 100 light-years ahead of it before it would encounter a small planet inhabited by curious creatures.

It arrives from the south. The date: Monday, September 14, 2015. The time: 09:50:45 Universal Time. For a fraction of a second, the

Earth is stretched and squeezed by a tenth of a quintillionth of a percent—one part in 10^{21}. Everything on the planet expands and contracts with it. That includes the Laser Interferometer Gravitational-Wave Observatory in Livingston, Louisiana. And, seven milliseconds later, the identical LIGO detector at Hanford, Washington.

Very soon, everything calms down again. The gravitational wave continues on its journey into the far reaches of outer space. After 1.3 seconds, it crosses the orbit of the Moon. Within a few hours, it will have left the solar system, still gently kneading everything that lies in its path.

———

Monday, September 14, 2015, is a day like many others. In London, the parents of singer-songwriter Amy Winehouse probably mourn their talented daughter, who would have turned thirty-two today had she not committed suicide just over four years ago. Space scientists with a feel for history recall the Soviet space probe Luna 1, which crash-landed on the Moon fifty-six years ago today, as the first man-made object on another celestial body. For most people, however, it's just a run-of-the-mill day. Nothing special.

That morning, postdoctoral researcher Marco Drago sits alone in his office at the Albert Einstein Institute in Hannover, Germany. He studied physics in Padua, Italy—the hometown of Galileo Galilei, who was one of the first to study gravity. In his free time Drago plays Mozart and Beethoven on the piano. He has also published two fantasy novels in Italian. They're about dragons and a young boy named Marco—*drago* is Italian for "dragon."

Around 11:54 local time, an email pops up in Drago's inbox. It's an automated alert from a LIGO data pipeline: lots of cryptic numbers and automatically generated hyperlinks. Apparently the software has detected some kind of anomaly, just three minutes ago or so. Interesting.

Drago clicks on one of the hyperlinks. Graphs pop up on his computer screen, showing the detector output. He knows what those graphs should look like: wiggly lines, revealing the incredibly tiny motions of the fused-silica mirrors that are suspended at the ends of the interferometer arms. Seismic noise, really. Even LIGO isn't able to hold the mirrors completely still at the level of one ten-thousandth of an atomic nucleus.

But this time it's different. Yes, the noise is there. But superimposed on it is a much stronger signal: a sine wave, alternately going up and down. Stronger and stronger. Faster and faster. Until it quickly dies away, and background noise takes over again. Everything happens in just a tenth of a second or so. And it's visible not just in the Livingston output. Hanford has it, too—a few milliseconds later. Not just interesting, *very* interesting.

Drago walks into the office of his colleague Andrew Lundgren, one door down the hall. Lundgren has been working here longer than Drago has. He is more experienced. Together they take a look at the graphs. The wiggly lines look exactly like the simulations Drago and Lundgren know so well. The rise in frequency and amplitude—it's the characteristic "chirp" of a gravitational wave signal. *Could this be . . . ? No, c'mon.* The signal is unexpectedly strong. It's plainly visible; you don't even need dedicated analysis software to dig it out of the noise. There must be another explanation. *This couldn't possibly be a real . . . Or could it?*

Let's call the control rooms, says Lundgren. The two detectors are running in engineering mode. The official start of the first observing run isn't scheduled until Friday. People are still carrying out all kinds of tests. This might very well be an intentional "hardware injection," just to measure the system's response. *Yes, that must be it. Let's not get nervous.*

At Hanford it's 3:30 AM. No one answers the phone. The operator on duty, Nutsinee Kijbunchoo, has just stepped out of the control room and missed the call. In Livingston (5:30 AM), operator William Parker says he doesn't know of any hardware injections. Sure, over the past two weeks, LIGO scientists Anamaria Effler and Robert Schofield have been doing lots of diagnostic tests. But yesterday was their last day. They actually worked very late and hadn't left until 4:30 AM or so.

Now what?

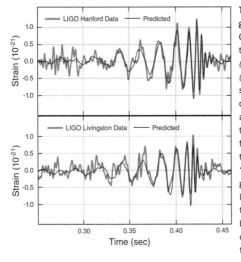

The first-ever captured gravitational-wave signal, GW150914, as observed by the LIGO detectors in Hanford *(top)* and Livingston, Louisiana *(bottom)*. These two graphs show the observed amplitude of the waves (strain, expressed as a fraction) as a function of time. Both amplitude and frequency increase over time; this is the characteristic "chirp" profile of a bona fide gravitational wave signal. Thick lines are actual measurements; the thin lines are "predictions" based on theoretical calculations of the merger of two black holes with masses of 36 and 29 solar masses.

Could there have been a hardware injection that no one knows about? A blind injection, issued by a secret team appointed by the LIGO Laboratory to keep everyone alert and to check the whole LIGO enterprise? Similar checks had been carried out before with Initial LIGO. But why now, before the first observing run has even really started? Moreover, Lundgren tells Drago that the complicated chain of events necessary for a proper blind injection is still in preparation.

At 12:54 PM, an hour after Drago first saw the alert email pop up, he sends out an email to the various teams of the LIGO Scientific Collaboration: the Burst Analysis Working Group, the Data Analysis Software Group, the Compact Binaries Coalescence Group, the Calibration team, the Detector Characterization people, the Commissioning and Observatory teams, the LIGO Open Science Center, and even the email list lsc-all@ligo.org.

Hi all,

cWB has put on gracedb a very interesting event in the last hour.
https://gracedb.ligo.org/events/view/G184098

Gibberish to anyone but a LIGO scientist. cWB is the coherent Wave Burst detection pipeline. GraceDB is the Gravitational Wave Candidate Event Database. (By the way, don't bother entering the link in your Internet browser—you need a LIGO login to access it.) A few more lines of hyperlinks follow. Marco concludes his email with a request for more information:

> It is not flag as an hardware injection, as we understand after some fast investigation. Someone can confirm that is not an hardware injection?
>
> Marco

It's still night or very early morning in the United States, so it will be hours before most American LIGO collaboration members read the message. But not Caltech's Stan Whitcomb. For some reason, he can't sleep. Around 4:00 AM, he gets out of bed and fires up his laptop to check on email. Drago's message has arrived minutes earlier. "Oh boy," Whitcomb mutters, "I'm going to be busy for the next few months."

Whitcomb has been at Caltech since 1980. He was one of the authors of the famous 1983 Blue Book—the first cost assessment for a LIGO-like interferometer. He closely collaborated with Ron Drever on the 40-meter Caltech prototype instrument. After working in industry for six years, he returned to LIGO in 1991, serving as the cochair for the site evaluation committee and, eventually, as LIGO's chief scientist.

Stan Whitcomb has recently announced that he will retire from Caltech on Tuesday, September 15. Not that he is planning to leave the field altogether. He has promised LIGO spokesperson Gabriela González that he will cochair the Detection Committee as soon as Advanced LIGO finds something interesting. That sounds like a quiet job—just a couple of meetings and teleconferences and a bunch of paperwork. But first he needs some well-deserved free time. He's planning to drive to Colorado on Wednesday to visit his mother.

And now this. A *real* chirp, not a simulation. And an exciting one, too. The short duration, the relatively low endpoint frequency—this can only be the collision of two pretty massive black holes. Lower-mass neutron stars would take longer to coalesce. Moreover, they would have a higher orbital frequency when they finally merge because they're much smaller in size. Whitcomb is almost immediately convinced that this is the real thing. LIGO has made its first detection. *Gotcha.*

Whitcomb can't go back to bed now, of course. Later that morning, when he and his wife are walking the dog together, he tells her, "I know I promised to spend more time at home after my retirement. But I'm afraid I'm going to be busy." He doesn't cancel his Colorado trip, but at his mother's place, he spends a couple of hours behind his computer every day.

Meanwhile, Gabriela González is upset. González is an Argentinian-born physicist at Louisiana State University in Baton Rouge. Since 2011, she has been the official spokesperson for the LIGO Scientific Collaboration, following in the footsteps of Rai Weiss of MIT, Peter Saulson of Syracuse University, and David Reitze of the University of Florida. Over the past months and weeks, she has been busy drafting and finalizing protocols and procedures. Apparently this Hannover postdoc isn't paying any attention to those protocols: Marco Drago has sent his message out to everyone, even to the "lsc-all" mailing list. Fortunately, that's a moderated list, and González happens to be the moderator, so that one is not going through. But she can't stop Drago's email from reaching all the other teams he addressed it to. By now, they're probably all talking to each other.

Of course, González is excited, too. At first, she thinks the powerful signal must surely be the result of some sort of test. But she quickly finds out none was going on at the time. She gets a text message from Mike Landry, the detection lead at the Hanford Observatory: "Gaby, have you authorized a blind injection?" Joe Giaime, observatory head at the Livingston facility, asks the same question. No, she hasn't. And indeed, *she* would be the person to have done so, in consultation with the project leadership. Easy checks prove there was no injection, at least not done in the usual way. So this must be a genuine gravitational wave—unless there's been a software glitch, an instrument anomaly, or a malicious hack. But that's for the Detection Committee to find out. It's all in the procedures.

Another procedure has not yet been put into place: the automated alert service to other observers. Some twenty ground-based and space-based observatories have arranged a special agreement with the LIGO Scientific Collaboration. As soon as a gravitational wave is detected, they plan to train their telescopes and instruments on the estimated point of origin (more on that in Chapter 14). The goal is to check if something is visible in X-rays or at optical or radio wavelengths. Any electromagnetic signal—like the high-energy radiation of some kind of explosive event—would travel at the same speed as the Einstein wave (the speed of light), so it should have arrived at the same time. But it might very well linger for longer than the gravitational wave did.

It's not yet certain that last night's signal is truly genuine. And even if it is, it's impossible to tell precisely where it was coming from. But González and Fulvio Ricci, the spokesperson of the European Virgo Collaboration, decide to send out a message to the counterpart search teams anyway with the coordinates of the huge swath of southern sky in which this gravitational wave—if it really is one—may have originated. If something is visible in the sky, it could fade away pretty soon, so people had better start looking as soon as they can, even though the positional accuracy is very low.

González's main worry, however, is secrecy. She needs to make sure no one outside the collaboration learns about this signal. Not yet. A key message in the detection protocol is that "confidentiality must be observed." After Joe Weber and after BICEP2, nobody wants to make yet another claim that has to be retracted at a later stage. It would be a terrible loss of face.

On Wednesday, September 16, González teams up with spokesperson Ricci and director Federico Ferrini of the Virgo Collaboration and with LIGO's executive director, David Reitze, and its deputy director, Albert Lazzarini. Together, they write an email to the thousand-plus members of what is loosely called the LIGO-Virgo Collaboration, or LVC. It reads as follows:

> Dear all,
>
> By now, many of you have heard about an interesting transient event candidate discovered in the ER8 data stream over the past weekend. . . . We have shared this with astronomy partners who can follow up this trigger. . . .
>
> **We want to remind everyone that we need to maintain strict confidentiality within the LVC,** specifically about this candidate and generally for all candidates and results. Conversations and communications about collaboration results with non-LVC members should *not* occur until results are made public. Leaks and rumors will only make our investigations that much more difficult.
>
> Some of you may be approached by your friends and colleagues about this event candidate and possible future candidates in O1. . . . Please do report any specific queries from people outside the LVC to the LSC and Virgo Spokespersons.
>
> Thanks!
>
> Gaby, Fulvio, Dave, Albert, Federico.

However, it's not easy to keep your mouth shut if your team has just made the discovery of the century. Marco Drago tells his parents in Italy. Stan Whitcomb tells his wife. Others tell their boyfriends or girlfriends. In Cambridge, Massachusetts, someone makes a typo in the address line of an email and inadvertently sends a message related to the discovery to the employees of MIT's financial department. Luckily, they know little physics. But it appears very likely that the news will leak, one way or the other.

And it does. Someone tells Lawrence Krauss, a theoretical physicist at Arizona State University in Tempe and the author of several popular science books. Krauss doesn't disclose the name of his source. It's not a member of the collaboration, but it *is* a distinguished, prize-winning experimental physicist—he's willing to say that much. On Friday, September 25, he posts the news on Twitter:

> Rumor of a gravitational wave detection at LIGO detector. Amazing if true. Will post details if it survives.

Krauss's tweet causes a buzz on social media. Gabriela González is becoming desperate. Journalists start to call her. Did LIGO really detect something? When? How? Why the secrecy? Has there been only one event? Will there be an official announcement? Later that day, she composes another email to the LIGO and Virgo Collaboration members.

> . . . Please *do not* provide any opinions or responses to these tweets, and of course refrain from divulging any information about this event. . . . Again, please do not participate in conversations in social media about this.
>
> Gaby
>
> PS: I am very disappointed this important news reached social media this fast - LSc has many members, but I sincerely thought we could count more with each other on keeping our science being done without these distractions.

She decides not to get in touch with Krauss. No one will. Plainly ignoring the rumors appears to be the best strategy for now. The official reply to media inquiries asks journalists to be patient: "We take months to analyze and understand foreground and background in our data, so we cannot say anything at this point."

Of course, science journalists also ask González about blind injections. Some of them remember the events of 2009 and 2011. Those were the days of Initial LIGO and Initial Virgo. Within the two teams, everyone knew about the possibility of blind injections. Two or three people high up in the collaboration had the authority to create a fake signal in the data stream of the interferometers. The goals were testing the efficiency of detection analysis software, checking if theorists would draw the right conclusions from the characteristics of the chirp signal, obtaining experience with writing professional publications, and finding out if anything needed to be changed in the procedures.

As soon as any potential gravitational-wave signal was detected and put up for further analysis, the blind injection team would seal an envelope containing the answer to the question of whether or not it was injected. Only when work on that particular signal was concluded would the envelope be opened to reveal the answer.

It certainly kept people busy. In the fall of 2007 and during most of 2008, LIGO scientists worked on a detailed analysis of a signal that was detected on September 22, 2007, fittingly called the Equinox Event. All three detectors—LIGO Hanford, LIGO Livingston, and Virgo—had registered a faint chirp within the noise. It looked like what you would expect from the spiral-in, collision, and merger of two neutron stars in a binary system.

However, subsequent analysis revealed that the Equinox Event wasn't really convincing enough to claim the detection of Einstein waves. The chances of the "signal" being a statistical quirk were just a bit too high. So in the fall of 2008, scientists agreed that the Equinox Event would not be considered a genuine gravitational-wave candidate.

Only in March 2009, when all analysis had been completed, was it revealed that the signal had been a blind injection. It was also revealed that there had been an earlier blind injection, just nine days before the Equinox Event, which apparently had been completely missed by the detection software. All in all, it was a very instructive experience.

A second famous blind injection, on September 16, 2010, was the Big Dog event. Big Dog was a much more conspicuous signal. Again, it was found in all three detectors. It looked like the chirp you would expect from a neutron star and a black hole smashing together. The tiny differences in arrival time at the three sites indicated that the collision had taken place somewhere in the constellation Canis Major— hence the name.

In this case, everything *did* look very convincing. Within months, scientists had completed the analysis and drafted a discovery paper to be submitted to *Physical Review Letters*. And, admirably, they had done so while being fully aware that Big Dog might very well be a fake signal. Only on March 14, 2011 (Albert Einstein's 132nd birthday), after the team had reached agreement on the final version of the paper, was the truth revealed at a meeting in Arcadia, California. Caltech's Jay Marx, LIGO's director at the time, opened the "envelope" (which was actually a memory stick containing a PowerPoint presentation), and an audience of some 350 collaboration members learned that they had been chasing a mirage. Champagne was still poured to celebrate the effort, though of course the *Physical Review Letters* paper was never submitted. Also, it was revealed that, once again, a second blind injection during the same observing run had not been found.

Both the Equinox Event and Big Dog had been injected in September. So it's little wonder that in September 2015, many scientists suspect they're being fooled again. But Gabriela González knows this is not the case. It cannot be. Of course, she's not sharing this information with inquisitive journalists. They are only told that fake signals have indeed been introduced in the past. But she *is* telling the collaboration members, all of whom should now realize that the signal detected on September 14 might be the real thing.

Might be, because now the hard work begins. The fact that the signal is not a blind injection doesn't necessarily mean it was produced by a genuine gravitational wave from space. There could be dozens of other possible causes. A software glitch, for instance. Detecting a tiny vibration in the mirrors at Hanford and Livingston depends on thousands of lines of computer code. Software is never bug-free, as every programmer knows. So this has to be checked in detail.

Or maybe the signal was due to an earthquake on the other side of the planet. Someone needs to check with the US Geological Survey. What about the atmospheric shock wave of a large meteorite or maybe even an impact in some uninhabited area? Someone else needs to run a check against infrasound recordings. Some strange phenomenon in

Earth's magnetic field may have been responsible. Satellite measurements collected by plasma physicists need to be reviewed. Sometimes, huge thunderstorms may even set up waves in the ionosphere, a layer of charged particles high up in Earth's atmosphere. Lots of natural phenomena could have triggered the delicate instruments. Everything needs to be checked and double-checked.

That's the task of the Detection Committee, chaired by LIGO's Stan Whitcomb and Frédérique Marion of the Virgo Collaboration. They know exactly what to do. The protocols are ready. The committee members—both in the United States and in Europe—all have their own individual tasks. Everyone is focusing on just a few potential issues. Checklists are ticked off; spreadsheets are filled out. Slowly but surely, each and every possible alternative explanation is ruled out. For instance, according to an international meteorology database of lightning strikes, there was an incredibly forceful bolt of lightning in Burkina Faso, in West Africa, around the time of the detection. Subsequent detailed analysis reveals that it cannot have affected the LIGO mirrors, though.

There are many other things to check. Sure, the chirp signal has been seen in both LIGO detectors, almost at the same time. The ability to obtain such coincident detections was the main reason to build two interferometers in the first place. But even then, you really need to rule out every possible *local* noise source, too. Simply put, a slammed door or a passing truck at both sites at the same time might be improbable, but it's not impossible. So that's another major task for the Detection Committee. Were there people in the interferometer tunnels at the time of the event? To find out, check all available logs, cameras, and microphones. Was there anything peculiar going on in the immediate neighborhood of the detectors? Recordings from all kinds of environmental sensors should provide the necessary information. Were there magnetic anomalies or other instrumental effects that might have disturbed the mirror suspension, the laser, or the photo detector? Everything is being monitored and logged, so you just have to weed through all the data to rule that out, too.

And then there's Stan Whitcomb's nightmare scenario: four grad students in a Livingston bar drinking too much beer and thinking of ways to hack the system, just for fun. After all, there are many extremely bright people in the collaboration. Maybe someone came up with an ingenious way of accessing the data stream or replacing an electronics board in one of the computer racks. Come to think of it, you can't even exclude a malicious act by a vengeful ex-employee. But in the end, even these remote possibilities are ruled out.

Whitcomb admits that absolute certainty is a rare commodity in science. The CIA or the North Korean secret police might be able to fool his Detection Committee. Or Tom Cruise, in a new *Mission: Impossible* movie. But not a prankster within the team or an invidious outsider. As one committee member says, if some individual has been successful in setting this all up, that would in itself be worthy of a Nobel Prize.

In the course of October and November, every possible cross-check is carried out: glitches involving nearby power lines, low-flying aircraft, mechanical wear in a vacuum pump, a technician who left a cell phone in the detector area. Nothing explains the September 14 signal. The Detection Committee even checks on the activities of the other teams and working groups in the collaboration to verify if they've done a good job. The whole process takes well into December.

By then, everyone is convinced. This is it. The first gravitational wave from deep space. And not the last one, by the way. On Monday, October 12, another plausible candidate is detected, albeit much less conspicuous than the first one. A third one, with a very strong level of confidence, is found on Saturday, December 26. Just over three weeks later, on January 19, Advanced LIGO's first observing run (O1) comes to an end. Finally, LIGO earns the *O* in its name. What for a long time was merely a technological miracle has now turned into an astronomical discovery machine. A real observatory. The ripples of spacetime are felt on Earth for the very first time. A century after Albert Einstein first aired the idea, his elusive waves are finally caught. The universe is broadcasting its secrets; elated scientists start to decode the message.

The chirp signal that first appeared on Marco Drago's computer screen on the morning of September 14, 2015, now gets a real name. No one doubts that this is a genuine gravitational wave anymore. It's called GW150914.

––––––––––

In October 2015, the LIGO Education and Public Outreach team is already starting to think about ways to break the news to the world. Not right now—certainly not before the discovery paper has been accepted for publication after an unbiased external review of the results, because no one wants a repetition of the BICEP2 experience—but perhaps within four months or so.

Also, any press conference should be meticulously prepared for and rehearsed to prevent potential miscommunication. The plan is to organize two simultaneous press conferences, one hosted by the National Science Foundation (NSF) in Washington, DC, and the other by Virgo at the European Gravitational Observatory in Italy. This has to be coordinated with many stakeholders. Maybe it's a good idea to bring in an outside expert, someone with lots of experience in space and astronomy communication.

High-energy physicist Fiona Harrison has a suggestion. Harrison is the chair of Caltech's Division of Physics, Mathematics, and Astronomy. She is also principal investigator for NASA's NuSTAR mission—the Nuclear Spectroscopic Telescope Array. In that capacity, she has worked with public information officer Whitney Clavin at the Jet Propulsion Laboratory (JPL) a few miles northwest of the Caltech campus, and she knows Whitney is a pro.

JPL is NASA's research and development center for planetary science missions and Earth observation. It is operated by Caltech; there are many ties between the two institutions. Clavin is of course very excited to hear about the news of the discovery. She's even more excited about the prospect of coordinating the media campaign.

At JPL, Clavin usually works closely with two graphic artists. Over the years, Robert Hurt and Tim Pyle have created hundreds of infographics, video animations, and beautiful artist's impressions on a wide variety of topics, ranging from extrasolar planets to infrared astronomy. Hurt is an infrared astronomer by training, and when Clavin tells him they will be doing all the graphics and artwork for LIGO's first detection of gravitational waves, he first shouts with joy and then is overwhelmed by emotion. Pyle, a professional graphic artist without a science degree, asks, "What are gravitational waves? And by the way, why is Robert crying?"

Like all LIGO collaboration members, they have to keep their mouths shut. Clavin can't disclose her whereabouts to anyone at JPL. Teleconferences take place behind closed doors. It's tough going—there are so many people to coordinate things with. And nothing—absolutely nothing—can be left to chance.

Then there's the process of training the scientists who will give presentations at the press conference in the United States: LIGO director David Reitze, collaboration spokesperson Gaby González, and founding fathers Rai Weiss and Kip Thorne. More emails, phone calls, and teleconferences ensue. Stay focused. Be concise. Keep your message simple and clear. Avoid scientific jargon. Use captivating metaphors. Clavin makes them practice again and again: three times on the phone, two times in person. Reitze enjoys it. Weiss and Thorne are grumpier; they sometimes feel it all takes the joy out of their presentation. In the end, however, everybody is happy with the result.

Ron Drever, LIGO's third initiator, will not be able to attend—at eighty-four, he suffers from dementia and lives in a nursing home in Glasgow. But France Córdova, the director of the NSF, will certainly be there. The press conference will take place at the National Press Club, close to the White House. Clavin makes sure the webcast has enough capacity. The press conference will be live-streamed on YouTube.

Picking a date turns out to be difficult, too. It has to fit in with everyone's busy schedules. The discovery paper must be accepted for publication in *Physical Review Letters*, as must two accompanying papers in another journal. In early January, the choice is made: Thursday, February 11, 2016, will be the day of the big announcement. Now it's just a matter of dotting the *i*'s and crossing the *t*'s. And of preventing any last-minute leaks, of course.

That turns out to be hard. On January 11, Arizona State University physicist Lawrence Krauss posts a second tweet:

> My earlier rumor about LIGO has been confirmed by independent sources. Stay tuned! Gravitational waves may have been discovered!! Exciting.

The tweet goes viral. LIGO scientists are pissed off. Krauss is being accused of being irresponsible and stealing the limelight. Personally, he believes that the role of social media is to connect the public at large directly with the scientific process. Also, his "movie trailer" for the announcement, as he calls it, may turn out to enhance media interest. On January 22, Krauss moderates a panel at his university titled "Einstein's Legacy: Celebrating 100 Years of General Relativity." LIGO's Kip Thorne is one of the panelists. Communication between the two theoretical physicists is awkward, to say the least.

Eight days before the press conference, there's another, much more specific, leak. An email by particle physicist Cliff Burgess of McMaster University in Hamilton, Ontario, Canada, finds its way to Twitter as an image attachment. It reads as follows:

> Hi all, the LIGO rumour seems real, and will apparently come out in nature Feb 11 (no doubt with press release), so keep your eyes out for it.
>
> Spies who have seen the paper say they have seen gravitational waves from a binary black hole merger. they claim that the two detectors detected it consistent with it moving at speed c given the distance between them, and quote an equivalent 5.1 sigma detection. the bh masses were 36 and 29 solar masses initially and 62 at the end. Apparently the signal is spectacular and they even see the ring-down to kerr at the end.
>
> Woohoo! (I hope)

Then, on February 8, LIGO announces the press conference. Apart from the regular media advisory channels, the announcement is also broadcast on Twitter:

> Announcement of LIGO press conference Feb 11 @ 10:30am EST!
> See http://bit.ly/1TLlihq to learn about #AdvancedLIGO &
> #GravitationalWaves!

That same day, British science writer Joshua Sokol posts a story on the website of *New Scientist* detailing his investigation of the online observing logs of the European Southern Observatory in Chile. He finds that follow-up observations for LIGO detections started on September 17, in a large region of southern sky. Another series of follow-up observations started on December 28, in the constellations Aries and Hydra. "LIGO may have been unbelievably lucky," Sokol writes.

By then, rumors are all over the place of a first signal on September 14, 2015, a second in late December, and maybe a third one in October. One day before the press conference, I do a Google search on "GW150914." After all, the Big Dog event was called GW100916 back in 2010. The search command yields exactly one hit: some obscure LIGO-project Internet page that shouldn't have been public. Apart from GW150914, it also makes mention of GW151012 and GW151226. So here are the three dates. No further details are available. A few hours later, the page cannot be reached anymore.

Finally, it's February 11. NSF director France Córdova gives a brief introduction. Córdova earned her PhD in physics at Caltech in 1978. Back then, Kip Thorne was on her thesis committee. LIGO was little more than "a glimmer in their eyes," she told me. Now, everything has changed. "Opening a new observational window will allow us to see our universe, and some of the most violent phenomena within it, in an entirely new way," she tells her audience. She also recalls that the initial funding of the LIGO project in 1992 was the largest investment NSF had ever made.

A brief kick-off video follows, with an upbeat musical score. It features the five panelists of the press conference, including Córdova.

"That's what scientific discovery is all about," she says. "You don't choose the simple things to do." Her comment is followed by a few numbers: 2 detectors, 1,000 scientists, 16 countries, 25 years. At the close of the video, Kip Thorne describes his feelings: "I looked at it, and I thought, 'My God.' This looks like it's it."

Then David Reitze walks to the rostrum. A TV screen shows a picture of two merging black holes. Reitze looks around the room, beaming, and says, "Ladies and gentlemen. We—have detected—gravitational waves. We did it!" Everyone bursts out in applause.

Reitze compares the discovery with Galileo Galilei opening up the field of observational astronomy. He characterizes the black hole collision as "mind-boggling." He compares LIGO's sensitivity to measuring the distance to the nearest star with the precision of the thickness of a human hair. "LIGO was truly a scientific moonshot," he says. "And we did it. We landed on the moon."

Gabriela González describes the actual detection. "This is the first of many detections to come," she assures her audience. She emphasizes it was the work of many people: "It takes a worldwide village." Next, Rai Weiss tells about gravitational waves. He demonstrates the squeezing and stretching of spacetime by pulling on a piece of plastic mesh in various directions. Weiss also explains how LIGO operates. "Had this technology been available to Albert Einstein, he would've invented LIGO," he quips. "He was smart enough; he knew enough physics."

Finally, Kip Thorne talks about his favorite topic, black holes. The merger of the two black holes, he explains, produced "a violent storm in the fabric of spacetime," which until now has always appeared to us as a glassy ocean. "The storm is brief but very powerful, momentarily unleashing fifty times as much power as all the stars in the universe together."

The webcast draws almost a hundred thousand continuous viewers; a few days later, half a million people have watched a replay. By the end of the February 11 press conference, at 11:15 AM, the news is all over the Internet. It may be only 2016, but people are already calling it the scientific discovery of the century.

LIGO director David Reitze *(left)* explains the discovery of gravitational waves at the February 11, 2016, press conference in Washington, DC. To his right are LIGO spokesperson Gabriela González and two of LIGO's "founding fathers," Rai Weiss and Kip Thorne.

The next morning, the *New York Times* opens with a science-fiction-like photo of a white-clad technician in one of the LIGO beam pipes. The headlines:

**WITH FAINT CHIRP, SCIENTISTS PROVE
EINSTEIN CORRECT**

A RIPPLE IN SPACE-TIME

**AN ECHO OF BLACK HOLES COLLIDING
A BILLION LIGHT-YEARS AWAY**

These may not be as captivating as the headlines were almost a hundred years ago when astronomers confirmed Einstein's prediction of the bending of starlight by the gravity of the Sun, but "Einstein would be beaming, wouldn't he?" as France Córdova puts it. That weekend, a photo circulates on the Internet, showing the giant statue of Albert Einstein at the campus of the Georgia Institute of Technology in Atlanta. Around his neck is a handwritten sign, saying, "Told you so."

In distant Germany, gravitational wave pioneer Heinz Billing, now 101 years old, has kept his 1989 promise to Karsten Danzmann: he lived until the waves were found. But Billing is now almost deaf and blind and suffers from severe memory loss. Like Drever, he lives in a nursing home. When he experiences a brief bright moment, his younger colleagues tell him about the LIGO discovery. "Yes yes, the gravitational waves," he replies in German. "I have forgotten so much." Billing died on January 4, 2017, at age 102.

It's unclear whether Ron Drever really understands and appreciates the news when it is brought to him by his relatives. However, his eyes gleam when he sees the chirp signal and watches the press conference. Within four months, Drever is declared corecipient—together with Rai Weiss and Kip Thorne—of four major science prizes: the Special Breakthrough Prize in Fundamental Physics, the Gruber Foundation Cosmology Prize, the Shaw Prize in Astronomy, and the Kavli Prize in Astrophysics. (The Breakthrough and Gruber prizes were shared by all members of the LIGO-Virgo Collaboration.) Meanwhile, Virgo's founding fathers, Adalberto Giazotto and Guido Pizzella, are honored with the 2016 Amaldi Medal, a European prize that is awarded every two years by the Italian Society of General Relativity and Gravitation.

Ron Drever died on March 7, 2017. But hardly anyone doubts that LIGO's remaining two founding fathers will someday receive the most coveted recognition in their field—a Nobel Prize in Physics.

(((12)))

Black Magic

No one has ever seen a black hole. Until recently, astronomers and physicists still debated their existence. Now they claim to have found two colliding holes at a distance of 1.3 billion light-years. Their circumstantial evidence? Tiny shivers of spacetime, no larger than one-thousandth of the diameter of a proton. For no longer than two-tenths of a second. Quite a leap of faith, you might say.

Astronomy isn't what it used to be, that much is clear. In the past, you just looked up in the sky to discover comets and exploding stars. That's what Danish astronomer Tycho Brahe did in the late sixteenth century. Later, a night at the telescope might reveal binary stars, dark markings on the surface of Mars, or the spiral structure of a faint nebula. What you saw is what you got.

Those days are gone. New discoveries—or claims of new discoveries—often are based on unconvincing-looking measurements and intensive data processing. A few photons here, an inconspicuous spectral feature there: it's all about statistical evidence and probability analysis. Always with the goal of extracting as much information as possible out of the available data.

Gravitational wave astronomy is no exception. Remember, the wiggly lines on Marco Drago's computer screen—the so-called chirp of GW150914—is really the only available piece of evidence we've got. A quick rise in frequency and amplitude, and then silence. What may have caused these tiny ripples? Run your data analysis software and out pops the answer: two coalescing black holes in the distant universe, which no one has really seen. It sounds like magic.

Still, theorists such as Kip Thorne are very confident about their claims. Since the signal has only been observed by two detectors, it's hard to say exactly from which direction it came. The distance, too, is not known precisely: it could be anything between 0.8 and 1.8 billion light-years or so. But the circumstances of the collision are much less uncertain.

In some remote galaxy, two black holes were orbiting each other. One was 36 times as massive as our Sun; the other weighed in at 29 solar masses—considerably more massive than astronomers had expected. (I'll return to that at the end of this chapter.) For black holes of this mass, the so-called event horizon—the spherical point-of-no-return "surface" of the black hole—has a diameter of a few hundred kilometers.

For many millions of years, the two black holes spiraled in toward each other ever so slowly as the emission of weak gravitational waves drew energy from the system, just as it does in the case of the Hulse-Taylor pulsar. As they got closer and closer, the black holes started to whirl around each other faster and faster. Stronger acceleration means larger-amplitude gravitational waves. A shorter orbital period means a correspondingly higher gravitational wave frequency.

Eventually, the two black holes approached within 350 kilometers or so of each other, racing around at more than half the speed of light. Then, within a fraction of a second, they merged together in a much more massive black hole, some 62 times the mass of the Sun. Now, every schoolkid knows that $36 + 29 = 65$, so what happened to the remaining three solar masses? It's been converted into energy ($E = mc^2$ all over again) and emitted in the form of a huge burst of Einstein waves.

Spacetime is extremely stiff, as I've noted before. But if you suddenly dump three solar masses' worth of energy into one particular

location, even spacetime can't help starting to tremble. Just beyond the event horizon of the final merged black hole, at a distance of a thousand kilometers or so, the gravitational waves stretch and compress the size of any object by as much as 1 percent for a very brief instant. That may not sound too dramatic, but it's enough to mess up the delicate chemical bonds of most molecules. You wouldn't survive. General relativity would kill you.

Seen from a somewhat safer distance, the coalescence of the two black holes must be a spectacular sight. At the press conference on February 11, 2016, Thorne showed a computer animation based on the same kind of scientific algorithms that had been used for the movie *Interstellar*. The animation shows the two black holes as ink-black circular disks silhouetted against a backdrop of stars. As they orbit each other, the light from background stars is bent this way and that by the strong gravity close to each black hole's event horizon. This gravitational lensing effect produces a hallucinating pattern of shifting and flickering stellar images. The two black holes spiral in toward each other, they merge, and then the resulting single black hole "rings down" like a struck gong, only much faster. At the end of the movie, everything is calm and quiet again. More than a billion years later, the spacetime ripples produced by the catastrophe reach Earth, albeit with an almost imperceptible amplitude.

So how can scientists be so sure that this is what happened? The movie looks convincing, but it's just an animation based on the equations of general relativity. How do Thorne and his colleagues know the masses of the two coalescing black holes, or the mass of the merger's end result? How do they know it really was a black hole merger instead of something else altogether? How can you tell on the basis of just two brief LIGO chirps?

This supercomputer simulation of two medium-mass black holes that are about to collide and merge is what a hypothetical local observer would have seen just before the production of gravitational wave signal GW150914. The swirling patterns around the black hole binary are caused by gravitational lensing of light from background stars by the strong spacetime curvature produced by the orbiting black holes.

Partly it's common sense and simple deduction. Einstein's theory tells you that a compact binary produces gravitational waves with a frequency that's twice the orbital frequency. Just before the merger, the observed wave frequency was something like 200 hertz, meaning that the two objects orbited each other about a hundred times per second. This alone tells you something about the huge masses and densities involved. The duration of the event is also revealing. For lower-mass objects, the spiraling-in would take longer. For smaller-diameter objects, the merger would occur at a higher orbital frequency. Finally, the Einstein wave frequency during the ring-down phase is determined by the mass of the final black hole.

Of course, for an accurate mass determination, you need a much more thorough analysis. And there's a big problem here: it's practically impossible to start with the observed wave pattern (the chirp) and then work backward to deduce the merger's characteristics from these data. Instead, theorists have to check the observations against many tens of thousands of pre-calculated wave patterns and look for the best possible match.

With fingerprints, it's the same. Each and every fingerprint is unique, so for every fingerprint found by a detective, there's only one matching person. But it's impossible to determine who that person is on the basis of just the fingerprint. Instead, you need a database of millions of fingerprints, within which you can search for a match.

That's why theorists have been busy calculating the expected gravitational wave patterns for a wide variety of merger events. In fact, for every conceivable merger event. What kind of Einstein waves do you expect from the collision of two neutron stars of 1.4 solar masses (like the two components of the Hulse-Taylor binary)? What if the two neutron stars are more massive? What if one is 50 percent more massive than the other? Or 40 percent, or 60 percent? What if it's one neutron star and one black hole? Or two black holes? What about tidal deformations? Eccentric orbits?

Different objects, different masses and mass ratios, a different viewing angle, the rotational state—for every possible variety, the resulting waveform can be calculated. Over the years, theorists have created a library of a few hundred thousand different waveforms. The characteristic chirp of GW150914 most closely matched the predicted waveform for two black holes with 36 and 29 times the mass of the Sun. So according to LIGO's "detectives," the "fingerprint" match indicates that this merger event must have been the "villain." It may sound like black magic, but it's serious science.

Those are no easy calculations, to be sure. The mathematics of general relativity is very complicated—that's why it took Einstein so long to formulate his ideas. For example, a black hole causes the surrounding spacetime to warp. The spacetime curvature represents a certain amount of energy. According to Einstein, energy is equivalent to mass. So the curvature's energy produces some additional curvature. Because of this so-called nonlinear behavior of general relativity, any calculation becomes very difficult and time-consuming.

Another complication is the notion of coordinate systems. In Isaac Newton's theory of universal gravitation, every event could be described with respect to absolute space and absolute time. Space and time constituted an invariant coordinate system for your calculations. However, in Einstein's theory of general relativity, nothing is absolute. Your coordinate system (spacetime) is itself affected by the very events you try to describe. In the case of a black hole, spacetime even gets hugely warped, drawn in, and gobbled up by the intense gravity of the black hole. You probably can imagine how hard it is to calculate an object's whereabouts if your coordinate system is ripped apart.

So calculating the expected waveforms from compact binary mergers is hard. Even for the simplest cases, it isn't something you could do with your pocket calculator, let alone on the back of an envelope. It took until the 1970s before mathematical physicists achieved their first successes. Today, most computational hurdles have been tackled. Still, it takes the brute processing power of a supercomputer to complete the calculations in a reasonable amount of time. So to build a library of a few hundred thousand different waveforms is quite an undertaking.

And of course, the Einstein wave library contains waveforms from events other than compact binary mergers. An asymmetric supernova explosion will produce a very different wave pattern. The same is true for a rapidly rotating neutron star with a tiny bump on its surface. Because of their high densities, neutron stars are expected to be the most perfect spheres in nature, but a "mountain" with an elevation of just 1 millimeter can produce observable gravitational waves. In all cases, the details can vary a lot, depending on the specific circumstances.

Anyway, given the close match of the observed wave pattern with the theoretical predictions, no one doubts that GW150914 was produced by the merger of two black holes that were 36 and 29 times as massive as the Sun. And since general relativity tells you the *original* amplitude of the resulting Einstein waves, it's pretty straightforward to calculate at what distance the collision occurred: just work back from the *observed* amplitude.

Likewise, the waveform of the second detection (GW151226) matched the prediction for the coalescence of two black holes that weighed in at 14.2 and 7.5 solar masses. This merger occurred at a slightly larger distance of 1.4 billion light-years. For obvious reasons, the analysis of this event didn't really start before February 11, 2016—the LIGO and Virgo scientists were just too busy with preparing for their first big announcement. Gabriela González, Fulvio Ricci, and David Reitze presented the GW151226 results on Wednesday, June 15, in a press conference at the 228th meeting of the American Astronomical Society in San Diego, California.

Because of the smaller masses involved, the spiral-in phase of the second event occurred at a slower pace. The observed chirp lasted for more than a full second, as compared to two-tenths of a second in the case of GW150914. The number of observed wave cycles was correspondingly larger: 54 cycles (corresponding to 27 orbits), as opposed to just 10 cycles (5 orbits) in the first event. And again, the final black hole weighed less than the sum of the two original ones: 20.8 solar masses. So in this case, 0.9 solar mass's worth of energy had been converted into gravitational waves.

What about the third detected signal, on October 12, 2015? According to the team, it may have been produced by the merger of two black holes at 23 and 13 solar masses and at a distance of more than three billion light-years. But the statistical significance of the detection was much lower than in the other two cases. Given the typical fluctuations in the background noise of the detectors, there is an estimated 1 percent chance that the event was not a real gravitational wave. For that reason alone, it didn't receive an official "GW" designation. Instead, it is now officially known as LVT151012, where "LVT" stands for LIGO-Virgo Trigger. Which is not to say that most collaboration members don't regard it as a genuine detection, albeit less convincing at a "mere" 99 percent confidence level.

So the waveforms of the first LIGO detections all point to coalescing black holes. According to some scientists, the gravitational-wave observations even constitute the first direct proof of the existence of black holes. Indeed, since by definition black holes do not emit any form of light (or other electromagnetic radiation), you cannot observe them directly—unless you "feel" the tiny vibrations they produce in the fabric of spacetime, of course. The only way for black holes to communicate with the rest of the universe directly is through gravity; the only language they speak is the language of gravitational waves. All other existing evidence for their existence is circumstantial and indirect.

———

The concept of black holes is actually much older than Einstein's theory of general relativity. Back in 1783, English clergyman and geologist John Michell was the first one to come up with the idea. This was just over half a century after Isaac Newton's death. The theory of universal gravitation was well known and appeared to be firmly established. Michell knew that every celestial body has a so-called escape velocity—the velocity you need to completely escape from the body's gravitational grip. For instance, the escape velocity of the Earth is known to be 11.2 kilometers per second; the escape velocity of the Sun is 617.5 kilometers per second.

What if the Sun was even more massive than it is? thought Michell. Its escape velocity would be higher still, of course. If a star is large and massive enough, the escape velocity might soar to 300,000 kilometers per second—the velocity of light. And what happens if light cannot escape from a star?

In a paper in the *Philosophical Transactions of the Royal Society of London,* Michell provided the answer: "If there should really exist in nature any bodies, whose density is not less than that of the sun, and whose diameters are more than 500 times the diameter of the sun, since their light could not arrive at us . . . of the existence of [these] bodies . . . we could have no information from sight." In other words, if light cannot escape their gravity, they're invisible to us. Michell didn't call those bodies "black holes," though. He called them "dark stars."

Of course, Michell's dark stars had nothing to do with curved spacetime—that concept didn't exist back in 1783. Also, eighteenth-century scientists weren't yet aware of the fact that the velocity of light is the highest possible velocity in nature. So Michell's hypothetical dark stars were not considered to be objects from which nothing could ever escape, like black holes. Even though *light* could not escape from a dark star, a spaceship conceivably could: just fire up your engines long enough (except that there were no spaceships back in 1783, of course).

The current concept of black holes dates back to early 1916. That was only a few months after Albert Einstein first presented his theory of general relativity. You may remember Einstein's field equations (immortalized on the eastern wall of Museum Boerhaave in Leiden). They turn out to permit the existence of regions in space where gravity is strong enough to curve spacetime back on itself, so to speak. This solution to the field equations was discovered independently by two brilliant scientists. The first was forty-two-year-old German physicist and astronomer Karl Schwarzschild. The second was Dutch mathematical physicist Johannes Droste, who was a twenty-nine-year-old graduate student of Hendrik Lorentz at the time.

At the start of World War I, in 1914, Schwarzschild had joined the German army. In the winter of 1915 / 1916, he was fighting both Russian soldiers on the eastern front and a rare blistering skin disease that may have led to his death in May 1916. In between, he found the time and mental focus to write three scientific papers, including one on what we now would call black holes. He exchanged letters about his finds with Einstein in Berlin as well. Droste's derivation, also much admired by Einstein, was more elegant but wasn't published until 1917.

Anyway, it was clear that a strong enough point-like gravitational field would exhibit a few strange properties. First, within a certain distance (now called the Schwarzschild radius), spacetime is curved so strongly that every possible motion, in whichever direction, ends up closer to the center than where it started. That's another way of saying that nothing can ever escape the region within this Schwarzschild radius, be it an elementary particle, a spaceship, or a ray of light. Second, the gravitational redshift at the Schwarzschild radius is incredibly strong. So strong, in fact, that time not only markedly slows down but even comes to a complete stop—at least as seen by an outside observer. Third, any matter that crosses the Schwarzschild radius (also known as the event horizon) ends up in the very center with an infinite density at a mathematical point of zero dimensions. At least, that's what the equations seem to indicate—probably a sign that our understanding of what goes on within a black hole is very incomplete.

Little wonder that most physicists, including Einstein himself, thought of the "Schwarzschild metric" as a funny mathematical quirk of general relativity. Something so weird can't possibly be part of our physical reality, right? After all, "permitted by general relativity" is not necessarily the same as "existing in nature."

But in 1934, Walter Baade and Fritz Zwicky predicted the existence of neutron stars, as we saw in Chapter 6. A neutron star is the collapsed core of a massive star that ended its life in a catastrophic supernova explosion. Recall that it consists of neutrons (uncharged nuclear particles) tightly packed together. In fact, you could describe a neutron star as an atomic nucleus the size of a city, and it certainly has the same incredibly high density as an atomic nucleus.

Five years after Baade and Zwicky's prediction, theoretical physicist Robert Oppenheimer—who would later become known as the "father of the atomic bomb"—argued that neutron stars cannot hold up against gravity if they're too massive. A neutron star with a mass of about three times the mass of the Sun would collapse even further. There just isn't any known law of physics to prevent that from happening. In fact, as far as theorists can tell, the gravitational collapse would never stop. According to calculations by Oppenheimer and his colleague Harlan Snyder, matter would simply be compressed to higher and higher densities, the end result being a region in space where gravity is so incredibly strong that nothing can escape.

But this is precisely what Schwarzschild and Droste had described in 1916: infinite densities, extreme spacetime curvature, trapped light, and a point-of-no-return "surface" where time appears to stand still. Oppenheimer and Snyder called these objects "frozen stars." The term "black hole" wasn't used until the 1960s. It first appeared in a news story by American journalist Ann Ewing in 1964 and was reintroduced by John Archibald Wheeler in 1967, half a century after the publications by Schwarzschild and Droste.

By then, astrophysicists couldn't really ignore black holes anymore. If the core of a massive star would collapse into a neutron star (after the star went supernova), then the core of a *very* massive star would collapse into a black hole. It was as simple as that. Nevertheless, many people remained skeptical about the existence of these enigmatic objects. It all sounded a bit too crazy. Moreover, if light cannot escape from a black hole, there's no way to observationally prove their existence, right?

Well, another half century has passed, and things have changed a lot. Over the past decades, astronomers have found loads of indirect evidence for the existence of black holes. A black hole itself is indeed invisible by definition. But what we *can* observe is the black hole's influence on its surroundings. It's a bit like the Invisible Man: you can't see him, but he leaves footprints in the yard, and when he sits on your bed, the sheets will get wrinkled.

Here's one way a black hole may betray its presence. Imagine a binary system consisting of two massive stars. The heftiest one evolves fastest, as we saw in Chapter 5. It goes supernova; its core collapses into a black hole. At a later stage, the second star starts to swell up into a giant.

The orbiting black hole sucks in the outer gas layers of the inflated star. Before plummeting into the black hole, the gas first accumulates in a thin, rotating disk surrounding the hole. This so-called accretion disk gets enormously hot and starts to emit X-rays.

This is exactly what scientists found in 1971. A luminous source of X-rays in the constellation Cygnus, the Swan—known as Cygnus X-1—turned out to coincide with a supergiant star. Doppler measurements revealed that the star is in a 5.6-day orbit around an object that's more than ten times as massive as the Sun. This massive companion object can't be a normal star or it would be visible in a telescope. It can't be a neutron star, either, since neutron stars cannot be more massive than three Suns or so. Moreover, the observed X-rays indicate that the massive companion somehow heats up gas to temperatures of many millions of degrees. The only possible explanation is a black hole, surrounded by a searing-hot accretion disk.

Many X-ray binaries are now believed to harbor black holes. Since they are the leftovers of exploding stars, they're called stellar-mass black holes, or just stellar black holes for short. In addition, astronomers have discovered much larger black holes in the cores of galaxies. These supermassive black holes can be anything between a few million and many billions of solar masses. In most cases, they reveal their presence by emitting copious amounts of high-energy radiation. They also blast powerful jets of charged particles into space. Such "active" galaxy cores are known as quasars. The energetic radiation originates in the accretion disk of the black hole. The jets are probably produced by powerful magnetic fields, although their origin is still something of a mystery.

Supermassive black holes also betray their existence by affecting the motions of stars in a galaxy's core. The velocity distribution of stars in the innermost region of a galaxy may indicate the presence of a very massive, very compact object at the center. In 1984, velocity measurements in the core of M32 (a small companion of the nearby Andromeda galaxy) led to the very first discovery of a supermassive black hole. In the case of our own Milky Way galaxy, astronomers have even observed individual stars swirling around an unseen object that weighs in at about 4 million solar masses. This can only be a supermassive black hole—there's just no viable alternative explanation.

Thanks to an ever-increasing amount of circumstantial evidence, black holes have gradually left the dark shacks of speculation and science fiction and entered the palace of accepted astrophysical reality. Even so, the detection of gravitational waves from two colliding black holes was seen as a welcome confirmation of their existence. Here, for the first time ever, was nature's clear message that black holes—or dark stars, Schwarzschild metrics, frozen stars, or whatever you want to call them—are part and parcel of our universe.

It was a powerful message, too. During the merger that produced GW150914, no less than three solar masses' worth of energy was released in the form of Einstein waves in a tiny fraction of a second. In fact, the black hole collision was one of the most powerful events ever observed in the universe.

Before impressing you with yet another round of astronomically large numbers, I first need to answer a question that may have been bothering you for a while. If black holes are regions of spacetime from which nothing can escape, how can they lose mass? Originally, the two black holes were 36 and 29 times more massive than the Sun. After the merger, however, a 62-solar-mass black hole remained. How could three solar masses have escaped the gravitational grip of the two black holes?

The straightforward answer is that it didn't happen. The merging black holes didn't magically spew out matter. In fact, it's a bit misleading to say that they contain matter at all. Whichever way a black hole forms, the matter that goes in is crushed out of existence in the central dimensionless point with infinite density—what physicists call the "singularity" of the black hole. What's physically left is the strong spacetime curvature. If an astronomer is talking about the mass of a black hole, she is referring not to a certain amount of matter but to a certain amount of spacetime curvature—one of the few observable properties of any black hole.

So here's what happened 1.3 billion years ago in that anonymous remote galaxy. Two spacetime "vortices," each with its own amount of curvature, were caught up in a violent spacetime "storm," where they merged into one larger "tornado." Most of the total available curvature (almost 95 percent) was used to form the resulting single black hole. Just under 5 percent (corresponding to three solar masses) was converted into gravitational waves.

If you enter three solar masses (6×10^{30} kilograms) and the square of the speed of light (9×10^{16} m^2/s^2) into Einstein's famous formula $E = mc^2$, you end up with an amount of energy equal to 5.4×10^{47} joules. That's 16 quadrillion times the total energy output of the Sun *per day*. Since this unimaginably large amount of energy was released during 15 milliseconds or so, the peak power output reached an incredible 3.6×10^{49} watts—almost ten times the combined radiation output of all stars and galaxies in the observable universe.

When Bruce Allen, the managing director of the Albert Einstein Institute in Hannover, tried to explain his excitement about GW150914 to his sons, Martin, then twelve years old, and Daniel, then fifteen, they weren't too impressed at first. Then Allen did some quick back-of-the-envelope calculations to compare the event's energy output to the devastating power of the Death Star, the "ultimate weapon" of the Galactic Empire in the *Star Wars* movies. "The Death Star is like a child's toy compared to this black hole collision," he told the boys. "The amount of energy released during the merger is more than enough to completely vaporize every single planet in every solar system in a hundred galaxies as large as the Milky Way." Now *that* finally made them go, "Wow, cool."

The other takeaway point here is that the collision and merger of two black holes is really an extreme-gravity event. In Chapter 3, we saw how physicists carry out all kinds of experiments to check on the predictions of Albert Einstein's theory of general relativity. But relativistic effects become important only in very strong gravitational fields (or at velocities close to the speed of light). Of course, it can be revealing to fly an atomic clock around the world in an airplane to measure the drift of a gyroscope in Earth orbit, or to time the delay of a space probe's radio signal when it disappears behind the Sun. But these are all low-gravity experiments. Even a binary neutron star is a "weak-field environment," at least as far as general relativity is concerned.

However, observing what happens at the event horizon of a black hole is a completely different story. It enables you to test Einstein's theory in a strong-field environment. And that's precisely where physicists expect possible deviations from general relativity's predictions. That's one reason why they're so excited about the prospects of gravitational wave astronomy. Spacetime ripples from colliding black holes provide you with a valuable probe of some of the most extreme environments in the universe. And that's where you want to carry out your tests; that's where you want to put Einstein on trial.

As discussed before, physicists think it's unlikely that general relativity is the final word on gravity. The theory is incompatible with quantum mechanics—that other giant pillar of twentieth-century physics. To make our description of gravity compatible with the tremendously successful description of the other forces of nature—and of all the particles that we know of—at least one of the two theories has to be adapted one way or another. The right path to the long-sought "theory of everything" is unknown, but maybe there's a helpful signpost close to the edge of a black hole. Studying Einstein waves from black hole collisions might reveal the signpost and help physicists to further improve their understanding of nature's most fundamental properties.

By the way, there's another possibility for studying general relativity in the immediate surroundings of a black hole. Radio astronomers, including Heino Falcke of Radboud University in Nijmegen, the Netherlands, and Shep Doeleman of MIT in Cambridge, Massachusetts, are linking up giant millimeter-wave radio telescopes in various continents. In doing so, they hope to create what they call the Event Horizon Telescope. The Event Horizon Telescope would have the sharpest vision of any astronomical observatory in history. The plan is to train the telescope at the supermassive black hole in the core of the Milky Way galaxy. Despite a distance of 27,000 light-years, it should be possible to see the black hole's event horizon silhouetted against the bright background of stars and glowing gas clouds. The view would more or less resemble the ink-black circular disks in the movie Kip Thorne showed during the LIGO press conference. The precise appearance of the black hole in the images would be compared to predictions from general relativity. Possible deviations might point the way to new physics.

———

New physics may still be a dream of the future. But the first detections of gravitational waves have already brought new *astro*physics. In fact, one of the GW150914 papers published on February 11, 2016, was completely devoted to the astrophysical implications of the discovery. Surprisingly, that very first event already provided new, important insights into the evolution of massive stars.

Before Advanced LIGO (aLIGO) came online, many collaboration members expected that the interferometer would mainly find neutron star collisions. In fact, the distance out to which neutron star mergers could be detected had become the standard way of quantifying an interferometer's sensitivity. For iLIGO and Initial Virgo, for instance, this "reach" was something between 50 and 65 million light-years; during the first observing run of aLIGO—working at one-third of its final design sensitivity—it was already 200 million light-years.

Sure, astrophysicists also expected black hole collisions to occur. If a pair of orbiting neutron stars spirals in toward each other, so will a pair of orbiting black holes. And yes, black hole collisions would be detectable at much larger distances: because of the larger masses involved, the amplitude of the resulting Einstein waves is also much larger. That's why GW150914 could be detected here on Earth despite its being 1.3 billion light-years away.

But nobody knew the number of binary black holes out there—not a single one had been discovered so far. Therefore, nobody knew how often you might expect collisions and mergers to occur. Estimates varied by many orders of magnitude. Binary neutron stars, in contrast, *have* been discovered in our own Milky Way galaxy; the Hulse-Taylor system was the first. Using a combination of statistics and some scientific guesswork, it wasn't too difficult to make a very rough estimate of the number of collision events that an interferometer such as LIGO might be able to detect. For iLIGO, the answer would be about one per decade or so; for aLIGO, just a few per year. (Remember that a threefold increase in sensitivity translates into a threefold increase in "reach," from 65 to 200 million light-years. But that corresponds to a volume of space that's 27 times larger, so the expected detection rate is also 27 times larger.)

So for neutron star mergers, scientists more or less knew the expected detection rate. That may have been the main reason why they believed that those events would be the first to be picked up by the advanced detectors. To physicists without enough astronomy background, it came as a surprise that the 2015 detections were actually of colliding black holes. Others, like Caltech's Stan Whitcomb, had always believed that black hole mergers would dominate LIGO's detections. Whitcomb's argument: they may very well be much less common, but you can "see" them out to much larger distances. And Kip Thorne, in his 1994 book *Black Holes and Time Warps,* even described a "future" scenario that is eerily reminiscent of what happened in September 2015:

> From the details of the waveforms, the computer extracts not only the history of the inspiral, coalescence, and ringdown; it also extracts the masses and spin rates of the initial holes and the final hole. The initial holes each weighed 25 times what the Sun weighs, and were slowly spinning. The final hole weighs 46 times what the Sun weighs and is spinning at 97 percent of the maximum allowed rate. Four solar masses' worth of energy ($2 \times 25 - 46 = 4$) were converted into ripples of curvature and carried away by the waves.

Pretty close!

By the way, in the case of GW150914, little information could be gained about the spin rates of the individual black holes. However, the data revealed that the final merged 62-solar-mass black hole was spinning at 67 percent of the maximum allowed rate. For GW151226 it was found that at least one of the two merging black holes was spinning at more than 20 percent of the maximum spin rate, while the final 21-solar-mass black hole was spinning at 74 percent of its maximum allowed rate. (Since black holes have no surface, it doesn't make sense to express a spin rate in revolutions per second, or to ask for a rotational velocity in kilometers per second. The maximum allowed spin rate of a black hole—or, more precisely, its maximum allowed angular momentum—is the value for which an object just outside the event horizon would whirl around at light speed.)

Given his visionary "prediction," Thorne may not have been too surprised by the discovery of black holes 36 and 29 times the mass of the Sun. But many astronomers were. Merging black holes was one thing; black holes as massive as those two was something else altogether. Sure, the black holes in the cores of galaxies are enormously more massive, but they have a very different formation history (more on that in Chapter 12). Black holes in a binary system, however, as mentioned before, are so-called stellar-mass black holes: they are the end products of the evolution of massive stars. And few astrophysicists could think of ways to create such heavyweights.

Naively, you might think that starting out with an extremely massive star will automatically leave you with a pretty massive black hole. But there are a couple of caveats here. First of all, you can't create a star as massive as you'd like. A huge cloud of gas that is contracting under its own weight will become hot and start to radiate, preventing more gas from raining down on the forming star. The presence of a sprinkling of heavy elements in the gas cloud will only enhance this effect. As a result, stars usually can't grow much more massive than about a hundred times the mass of the Sun.

Wouldn't that be enough to produce a 36-solar-mass black hole? Well, not really. During their brief lives, extremely massive stars lose most of their outer layers to space through powerful stellar winds. Again, such winds are stronger if the star contains a small amount of elements heavier than hydrogen and helium. So at the very end of its brief life, our 100-solar-mass star may well have lost more than half of its weight. A substantial part of what's left will be ejected during the final supernova explosion. The remaining stellar core that collapses into a black hole is expected to weigh in at no more than 10 to 15 solar masses.

So now you'll understand why astronomers were so excited about the very first LIGO detection. It constituted the first direct evidence of the existence of black holes. It also made clear that *binary* black holes do indeed exist—remember that no one had ever discovered such a system before. Third, it revealed that nature is capable of forming stellar black holes that are much more massive than the generally accepted limit of about 10 solar masses.

Gijs Nelemans of Radboud University was one of the two coordinating editors of the GW150914 astrophysics paper in *Astrophysical Journal Letters*. (Nelemans is the grandson of Anton Pannekoek, a contemporary of Albert Einstein and the founding father of Dutch astrophysics—the astronomical institute at the University of Amsterdam is named after Pannekoek.) According to Nelemans, GW150914 was a generous gift of nature. Not only was it the first Einstein wave ever to be detected, it also provided important new information on the birth and evolution of massive stars.

Nelemans and his fellow authors believe that the progenitors of the merging black holes must have contained very few heavy elements. That would have decreased their mass loss through stellar winds. Also, if they were born from a relatively "pure" cloud of interstellar gas, with negligible quantities of elements heavier than hydrogen and helium, they may have started out as true stellar heavyweights. Slightly tweaking the generally accepted astrophysical wisdom might explain the formation of black holes that are dozens of times as massive as the Sun.

Many questions remain unanswered for now, including the precise formation history of the binary black hole. Did it start out as a pair of extremely massive stars? Or did the black holes partner with each other long after their formation? According to some theories, black holes that weigh in at a few tens of solar masses may even date back to the very early days of the universe. Whichever formation scenario is right, additional discoveries of coalescing stellar black holes will certainly shed new light on the birth, evolution, and death of the most massive stars in the universe. In addition, astronomers expect to learn more about the properties of black holes in general.

And what about the supermassive black holes in the cores of distant galaxies? What can gravitational waves tell us about those cosmic monsters? Quite a lot, it turns out. But not through laser interferometers like LIGO and Virgo. Instead, we need to use the cosmos itself as our detector. Time to return to pulsars again.

(((13)))

Nanoscience

Parkes is a small town in pastoral New South Wales, Australia, some five hours west of Sydney. Founded in 1853, it is named after Henry Parkes, who is often called one of the fathers of the Australian federation. From the unassuming city center, it's just another twenty-minute drive to "the Dish." Head out of town northward on Newell Highway, then take a right at Telescope Road. Within minutes, you've reached the giant radio telescope.

Construction of the Dish—the unofficial name of the 64-meter Parkes radio telescope—was completed in 1961. Back then, radio astronomy was still in its infancy. But apart from studying radio waves from the sky, the instrument also played a role in tracking spacecraft. During the 1960s, it picked up transmissions from NASA's interplanetary probes Mariner 2 and Mariner 4. And in July 1969, the Dish was instrumental in receiving live TV images from the historic Apollo 11 lunar landing. (Be aware, though, that the 2000 movie *The Dish* by Australian director Rob Sitch is a strongly fictionalized comedy, not a documentary.)

To astronomers, the Parkes observatory is known mainly for its pulsar research. Almost half of all the known pulsars in our Milky Way galaxy have been discovered by this telescope. Yes, it's an old-fashioned

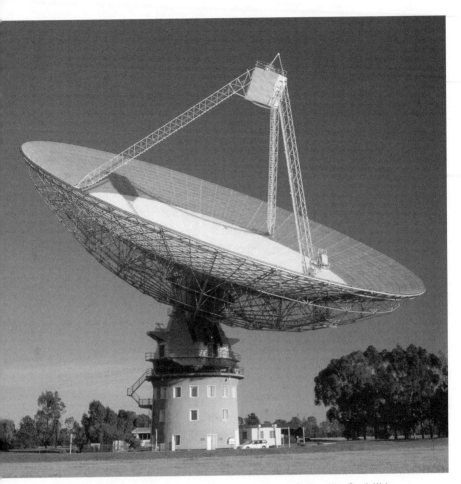

"The Dish" is the nickname of the 64-meter radio telescope in Parkes, New South Wales, Australia. The telescope has made vital contributions to pulsar astronomy.

instrument by today's standards, but pulsars are still observed almost daily. One of the research goals is the detection of gravitational waves through pulsar timing measurements.

As I explained in Chapter 6, a pulsar is a rapidly rotating neutron star that happens to have a favorable orientation in space (favorable for us, that is). As it spins around its axis, one of its lighthouse beams sweeps across Earth once per revolution. The result is an incessant series of brief pulses of radio waves at extremely regular intervals. Some pulsars are more accurate timekeepers than atomic clocks.

Thanks to this incredible regularity, measurements of pulse arrival times can yield all kinds of information about the pulsar's motion. That's how Joe Taylor and Joel Weisberg discovered the slow orbital decay of the first binary pulsar, PSR B1913+16. As you'll remember, this was the first convincing indirect evidence for the existence of gravitational waves.

But there's another, more direct way for a pulsar to betray the existence of our elusive spacetime ripples. Suppose a gravitational wave travels through the universe, alternately squeezing and stretching space itself. If the wavelength is long enough—meaning that the stretching and squeezing occurs very slowly—it should be possible to detect the effect in the pulse arrival times of a distant pulsar. The reason is as follows: as soon as the space between the Earth and the pulsar expands a little bit, the pulses take longer to arrive at our radio telescope. If it contracts a little bit, the pulses arrive a bit early.

Of course, a brief event like GW150914 cannot be observed in this way, since it would hardly affect a single pulse. But slow, continuous undulations of spacetime—not waves with a frequency of a few hundred hertz, but ones with a frequency of a few nanohertz, which is about a hundred billion times slower—might be observable. Such extremely low-frequency Einstein waves are indeed expected to exist. They should be generated by binary supermassive black holes at the cores of distant galaxies. You can't detect those nanohertz waves with laser interferometry. Instead, you need to use our own galaxy as a detector. You also need a lot of patience, as the Parkes radio astronomers and their international colleagues will confirm.

Soviet astrophysicist Mikhail Sazhin of the Sternberg Astronomical Institute in Moscow was the first to suggest the use of pulsars to directly detect nanohertz gravitational waves back in 1978. One year later, in the *Astrophysical Journal,* Yale University astronomer Steven Detweiler also described pulsar timing measurements as a means to search for gravitational waves. However, Detweiler concluded that the observations had to become much more precise for the technique to actually work.

Obviously, if you want to find low-frequency Einstein waves by looking for minute variations in pulse arrival times, what you need is an extremely regular pulsar. Moreover, ideally the pulses should be very brief, to enable the most accurate timing possible. A pulsar such as PSR B1919+21—the first one discovered by Jocelyn Bell in 1967— isn't very useful. Its pulses last for about 40 milliseconds. (They're pretty irregularly shaped, too, as fans of the British postpunk band Joy Division know—the cover of their 1979 debut album *Unknown Pleasures* famously features chart recordings of Jocelyn's pulsar.)

As luck would have it, a new, ideal type of pulsar was accidentally discovered in 1982. Don Backer and Shrinivas Kulkarni of the University of California at Berkeley studied a mysterious source of radio waves in the Milky Way known as 4C21.53. Astronomers had never found this radio source to be pulsating. But might it be flickering so unbelievably fast that the pulses just hadn't been detected before? Backer and Kulkarni decided to check. To their surprise, 4C21.53 indeed turned out to be a pulsar, and with an incredibly short rotation period of 1.5577 milliseconds. This huge ball of neutrons, some 50 percent more massive than the Sun and as large as a city, was spinning around at 642 revolutions per second.

Backer and Kulkarni had discovered the first millisecond pulsar. It is now known as PSR B1937+21, after its coordinates in the sky. It's not too far from where Jocelyn Bell had found the "Joy Division pulsar" fifteen years earlier, but at a much larger distance from the Earth.

Before long, radio astronomers had found other millisecond pulsars. Most of them are part of a binary system. Apparently, gas from the companion star has accreted on the compact neutron star. The inflow of gas has spun up the neutron star's rotation rate, just as a pinwheel is spinning faster and faster when blown on in the direction of rotation. Since millisecond pulsars are rotating so rapidly, their radio pulses last for only a tiny fraction of a second. Moreover, they turn out to be extremely stable.

One of the most famous millisecond pulsars is PSR B1257+12. It is located in the constellation Virgo at a distance of some 2,300 light-years. It was discovered in 1990 by Polish radio astronomer Aleksander Wolszczan, using the 305-meter Arecibo radio telescope—the same instrument that found the Hulse-Taylor pulsar in 1974. The pulse frequency was 161 hertz, corresponding to a spin period of 6.22 milliseconds—not particularly fast for a millisecond pulsar. But something else caught Wolszczan's attention: the pulse period wasn't perfectly constant.

In 1992, together with his American colleague Dale Frail, Wolszczan came up with an astounding explanation: *two* small objects were orbiting the pulsar, at periods of 66.54 and 98.21 days, and as a result, the pulsar displayed tiny periodic wobbles. Thanks to the Doppler effect, those minute motions betrayed themselves in the pulse arrival times. From the timing measurements, Wolszczan and Frail were able to deduce the masses of the pulsar's companions: 4.3 and 3.9 times the mass of the Earth. For the first time in history, astronomers had discovered planets orbiting a star other than our own Sun.

Two years later, a third planet was found in the data, just twice as massive as the Moon, in a 25.26-day orbit. In December 2015, the International Astronomical Union officially named the three planets Draugr, Poltergeist, and Phobetor after various ghost- and zombie-like mythological creatures. This choice of names referred to the fact that the three small objects are orbiting the mortal remains of a star that went supernova; in fact, the planets may have formed from the debris of the supernova that produced the pulsar. (The first planet orbiting a more or less Sun-like star wasn't discovered until 1995.)

What's important here is that the planets never would have been discovered if PSR B1257+12 hadn't been a millisecond pulsar. Fast rotation, clockwork precision, and extremely short pulse duration—that's what enabled the timing accuracy needed to find and study the slight variations in pulse frequency.

Over the past decades, almost 150 millisecond pulsars have been found in the Milky Way. Many of them reside in globular clusters—

huge spherical swarms of hundreds of thousands of stars. That makes sense: in the densely populated cores of globular clusters, pulsars have a bigger chance of ending up in a binary system and being spun up by their companion star. The large globular cluster 47 Tucanae, for instance, is home to at least twenty-two millisecond pulsars. Another cluster, called Terzan 5, contains no fewer than thirty-three of these rapidly spinning zombie stars.

One of the millisecond pulsars in Terzan 5 is known as PSR J1748–2446ad. It was discovered in 2005 by Canadian Dutch astronomer Jason Hessels. With its rotation period of 1.396 milliseconds, it's the fastest one known so far. Its spin rate is 716 revolutions per second—faster than your kitchen blender. The rotational velocity at the pulsar's equator is about 25 percent of the speed of light.

By the late 1980s, it had already become evident that millisecond pulsars would be the ideal objects to use as galactic probes for very-low-frequency Einstein waves. This was well before construction of LIGO got started. In fact, some pulsar astronomers believed they might be able to scoop the laser interferometrists in achieving the first direct detection of gravitational waves.

Berkeley radio astronomers Don Backer and Roger Foster set out the procedure in a 1990 paper in the *Astrophysical Journal,* "Constructing a Pulsar Timing Array." The plan was to keep an eye on a number of millisecond pulsars distributed over the sky—that's your array. When you're observing only one pulsar, you can never be sure that timing variations are really due to gravitational waves. But if you precisely measure the pulse arrival times for many millisecond pulsars for an extended period of time, you'll have enough data to single out slight deviations that can only be the result of a background of low-frequency gravitational waves. The longer you continue your timing measurements, the better your chances of success.

For their experiment, Backer and Foster used the National Radio Astronomy Observatory's 43-meter radio telescope in Green Bank, West Virginia. They collected data on three millisecond pulsars for two years or so. The first one was PSR B1937+21—the very first millisecond pulsar ever found, discovered by Backer and Kulkarni in 1982. The second one was PSR B1821–24, located in the globular cluster M28. The third was PSR B1620–26, found in another globular cluster, M4. (Interestingly, the timing measurements of this third object eventually revealed that it, too, was accompanied by a planet.)

Three pulsars and two years of data were not enough to detect gravitational waves. But at least they were a start. If astronomers could collect precise timing measurements of dozens of pulsars all over the sky for at least a decade or so, the nanohertz waves should show up. Time to get down to work.

————

Before we go on, I need to tell you a bit more about those nanohertz waves and where they come from. These are very strange waves indeed. As you will remember, the period of any wave is the inverse of its frequency. If a wave has a frequency of 100 hertz, it means that 100 wave crests (and troughs) are passing you by every second. So the wave's period (the time between two successive wave crests) is 1/100 of a second. Waves with a frequency of 1 hertz (one cycle per second) obviously have a period of 1 second.

So any wave phenomenon with a frequency of 1 nanohertz (one-billionth of a hertz) has a period of a billion seconds. That's more than thirty years! If a passing gravitational wave has a period of 1 nanohertz, space is slowly expanding a tiny bit for something like fifteen years before contracting again for another fifteen years. The *amount* of stretching and squeezing—the amplitude of the wave—may still be very small, on the order of one ten-trillionth of a percent. So we're trying to detect minute variations that play themselves out at a snail's pace.

Another thing to keep in mind about nanohertz gravitational waves is that they, too, propagate at the speed of light. If the wave's period is 30 years, its wavelength is 30 light-years. So if I'm talking about "slow" waves, I'm referring not to their actual velocity (which is the highest velocity permitted in nature) but to the long time it takes for them to make their presence felt.

What kind of cosmic events might produce such extremely-low-frequency spacetime ripples? Well, we've seen that gravitational waves are generated by orbiting bodies such as binary neutron stars and binary black holes. You also may remember that two wave cycles are being produced per orbit—if two black holes orbit each other 100 times per second (as they did in the case of GW150914 just before they collided and merged), the Einstein waves they produce have a frequency of 200 hertz. In other words, the wave's period is half the orbital period.

A gravitational wave with a frequency of 1 nanohertz has a period of about 30 years, as we just saw. So those waves should be produced by celestial bodies that orbit each other once every 60 years. But two neutron stars or two stellar-mass black holes in a 60-year orbit do not produce detectable gravitational waves—the masses and the accelerations are just too small. Remember that GW150914 became observable to LIGO only when the amplitude of the waves started to increase dramatically, just before the two black holes collided and merged.

For two objects in a 60-year binary orbit to produce gravitational waves at a detectable level, they have to be very, very massive. Think supermassive black holes at the cores of distant galaxies: two hungry black monsters, each millions of times more massive than the Sun, in a slow dance, orbiting each other once every six decades. Actually, they're engaged in a dance of death: just like their low-mass cousins, they're spiraling in toward each other, and will collide and merge in the distant future.

If binary supermassive black holes do exist in the universe, we would expect them to have a wide variety of orbital periods, of course, ranging from months to millennia. The Einstein waves they produce would span a correspondingly wide range of frequencies, from a tenth of a millihertz to 10 picohertz or so. Obviously, it's going to be tough to observe gravitational waves with periods of centuries. Their effects won't change much within a human lifetime, so there's little chance of detecting them. Moreover, for such long orbital periods, you need extremely large black hole masses to produce waves with sufficient amplitude. But pulsar timing arrays should be able to pick up waves with frequencies between, say, 1 and 10 nanohertz.

So do binary supermassive black holes exist? Yes, they do. As you read in Chapter 12, most galaxies have supermassive black holes at their cores. They probably formed many billions of years ago, together with the galaxies themselves. Details of their origin are still sketchy, but astronomers have detected quasars out to distances of well over 12 billion light-years. Quasars (short for quasi-stellar objects) are the luminous, energetic nuclei of galaxies that are "powered" by extremely massive black holes. Seeing them at such large distances means they already existed when the universe was still young. Who knows, maybe the birth of *every* galaxy involved the formation of a supermassive black hole.

Now, if there are *single* supermassive black holes, there must also be *binary* supermassive black holes. That's because galaxies collide and merge over time. Even in an expanding universe, neighboring galaxies—for instance, in large galaxy clusters—feel each other's gravitational pull. They are drawn closer and closer together, and eventually they will coalesce into one larger galaxy. If both galaxies have a central supermassive black hole, the two holes will be drawn together as well, and they'll end up as a binary supermassive black hole at the core of the merged galaxy.

Astronomers are witnessing galaxy mergers all over the place. Of course, those are happening much too slowly for us to see anything taking place in real time. Instead, we're presented with stills from cosmic collision movies, like short-exposure photographs of traffic accidents. Distorted spiral shapes, tidal tails of gas and stars, renewed star formation activity—each and every imaginable stage of a galactic collision has been found in the universe around us. Combine those observations with detailed computer simulations, and you get a pretty good picture of the whole process.

In fact, our own Milky Way galaxy is on a collision course with its nearest neighbor, the Andromeda galaxy. They're still 2.5 million light-years apart, but they're approaching each other at some 100 kilometers per second. A few billion years from now, the two majestic spiral galaxies will crash and merge into a humongous elliptical galaxy. And since they both harbor a supermassive black hole at the center, the resulting galaxy (called Milkomeda) will end up having a binary supermassive black hole at its core.

Binary supermassive black holes have even been observed out there, albeit indirectly. Periodic brightness variations and Doppler measurements of distant quasar-like objects, some 3.5 billion light-years away, are the evidence. Detailed observations and supporting computer modeling leave room for only one interpretation: two very massive black holes orbiting each other. At present, the black holes are still separated by trillions of kilometers (a substantial fraction of a light-year). They're expected to merge tens of thousands of years from now.

————

So you'd expect the universe to be flooded by very-low-frequency gravitational waves. They arrive from every possible direction in space. They have a pretty wide range of frequencies (but all more or less in the nanohertz regime). They also have very different amplitudes, depending on the black hole masses involved and, of course, on the distance the waves have traveled. Together, they continuously stretch and squeeze spacetime a tiny little bit, this way and that, very slowly. It's what astronomers call the gravitational wave background.

Here's a revealing analogy. Suppose you're in a tiny boat on a pretty calm ocean. It's not too difficult to observe small ripples on the surface. If someone throws a large stone in the water close to your boat, you will feel how your boat starts to bob a little bit. But it's much more difficult to detect very slow, continuous undulations of the water surface—waves with a larger amplitude, perhaps, but with a much lower frequency. How would you go about measuring this "wave background"?

The answer is actually quite simple: your "detector" is not your own boat, but those around you. The other boats floating on the ocean may, like yours, wobble a bit in response to small, fast waves, but if you observe them for a long time, those motions will get averaged out. The low-frequency waves, however, will cause the other boats to bob up and down very slowly. Measuring the long-term motions of a number of those boats will reveal the existence of slow undulations of the ocean surface. If you know the distances to the individual boats, and if you collect enough measurements, you may even be able to recognize a few discrete sources of low-frequency waves.

That's exactly how a pulsar timing array works. The ocean surface is spacetime. The surrounding boats are millisecond pulsars in our Milky Way galaxy. The pulsars are not bobbing up and down, of course (as I've said before, no single analogy is perfect). Instead, when a low-frequency gravitational wave passes, the space between the Earth and a particular pulsar is alternately stretched and squeezed—actually, the amount of space between the Earth and the pulsar is increasing and decreasing again, very slowly, by a very small amount. But if you keep track of the pulse arrival times for many years on end, the effect should eventually show up. Easy.

Well, it's not *that* easy. If both the Earth and the pulsar sat still in space, and if the pulsar was a truly perfect clock, then all variations in pulse arrival times would be due to Einstein waves. But things are much more complicated. First of all, pulsars are not perfect—nothing in nature is. Their rotation slows down, albeit very gradually. They also may exhibit "glitches"—sudden tiny changes in their rotation period. Glitches may be caused by "star quakes" on the neutron star's surface, or by interactions of the crust with the superfluid interior. If you don't measure those effects and correct for them, you'll never be able to spot a gravitational wave.

Moreover, millisecond pulsars are often part of a binary system. You need to correct for their orbital motion, which also affects the pulse arrival times. Likewise, corrections are needed for the motion of your radio telescope through space. The rotation of the Earth, Earth's orbital motion around the Sun, the small gravitational disturbances from other planets in the solar system, tidal effects, the Sun's motion through the Milky Way galaxy, even continental drift—everything has to be taken into account. The trick is to precisely model all possible influences and to subtract them from the measurements. Any remaining deviations from a completely steady stream of pulses might possibly be due to gravitational waves.

In principle, you could do this experiment with a single millisecond pulsar. But then you could never be sure that you're really measuring gravitational waves and not something else. That's why you need more pulsars, and the more the better. Preferably all over the sky, randomly distributed. You'll need to keep a very close watch on them for years—or, even better, decades. The longer you observe them, the more sensitive your experiment becomes. And knowing the distances to the pulsars helps a lot in analyzing the observations. Maybe you will find a couple of stronger-than-average nanohertz gravitational wave sources, from relatively nearby binary supermassive black holes, superimposed on the more chaotic background.

The great thing about pulsar timing arrays is that they're free. The Milky Way galaxy is full of high-precision clocks. No need to develop and build complicated and expensive laser interferometers. The only thing you need is a large enough radio telescope—an old existing one might do—and electronics to dig the pulsar signals out of the observations and to accurately measure the pulse arrival times. That's pretty complicated, but it won't necessarily cost you hundreds of millions of dollars. In a way, observing pulsar timing arrays is a poor man's way of hunting gravitational waves.

The hunt will require perseverance and patience, however. This is slow-paced science. If you start your project today, don't expect to have results any sooner than ten or fifteen years from now. At least, that's been the case so far for the Parkes Pulsar Timing Array (PPTA) project in Australia. It officially got going in 2004, but so far nothing definitive has been detected. So team leader George Hobbs of the Australia Telescope National Facility and his thirty-plus team members patiently keep on collecting more data to improve the experiment's sensitivity.

The PPTA project uses just one instrument—the 64-meter Dish at Parkes. (Note that the word *array* refers to an array of pulsars, not an array of telescopes.) In between other observing programs, the huge radio telescope is trained at some twenty millisecond pulsars to collect a few minutes' worth of timing measurements for each of them. If, say, the pulse frequency is 200 hertz, five minutes corresponds to 60,000 individual pulses. Each radio pulse may last for a tenth of a millisecond or so, and pulses may look different from each other. But averaging out more than 60,000 pulses enables you to determine the pulse period to a precision of something like 100 nanoseconds, or 1/10,000 of a millisecond.

A similar observing strategy is carried out in Europe. The European Pulsar Timing Array (EPTA) project, which started in 2006, is using five different radio observatories. One of them is the venerable 76-meter Lovell Telescope at Jodrell Bank Observatory in the United Kingdom. It has been observing pulsars since 1969, shortly after Jocelyn Bell's landmark discovery. An even bigger instrument is the 100-meter dish at Effelsberg, Germany. The Westerbork Synthesis Radio Telescope in the Netherlands—itself a linear array of fourteen 25-meter dish antennas—has been doing pulsar work since 1999. The fourth EPTA telescope is the huge Nançay Decimetric Radio Telescope in central France. Finally, in 2014, the recently completed 64-meter Sardinia Radio Telescope in Italy joined the collaboration.

Observing the same pulsars with three or more telescopes has one big advantage. If you're using a single telescope, a freak technical issue with the instrument might mess up your data without you ever finding out. With two telescopes, you would at least discover that something is amiss, because the two instruments would yield different results, but you wouldn't know in which one the problem occurred. With three, you're safe. The five European telescopes have very different designs, so combining the various datasets can be a complicated business. However, European pulsar astronomers have now standardized their pulsar timing instrumentation in order to get even better measurements.

Since 2007, two large American radio telescopes also officially work together in observing a pulsar timing array. They are the giant Arecibo dish in Puerto Rico and the 100-meter Green Bank Telescope in West Virginia. The North American Nanohertz Observatory for Gravitational Waves (NANOGrav), as the project is called, brings together a few dozen radio astronomers from fifteen universities and institutes. The three groups (PPTA, EPTA, and NANOGrav) work together in a loose collaboration called the International Pulsar Timing Array (IPTA).

Just a few years ago, radio astronomers still secretly hoped that they might be able to discover direct evidence of Einstein waves before the LIGO and Virgo physicists. In 2010 and 2011, the two laser interferometers shut down for their major upgrades. So far they hadn't found any gravitational waves, and the advanced detectors weren't expected to come online until 2015 and 2016, respectively. Meanwhile, pulsar observations were continuing relentlessly. In 2013, NANO-Grav's principal investigator, Xavier Siemens, and his colleagues even wrote an optimistic article in *Classical and Quantum Gravity* stating that "a detection is possible within a decade, and could occur as early as 2016."

Obviously, that didn't happen. The LIGO and Virgo collaborations took the world by surprise with their first batch of discoveries. On February 12, 2016, just one day after the GW150914 press conference, the following message was posted on the IPTA website:

> The International Pulsar Timing Array (IPTA) would like to congratulate our LIGO and Virgo colleagues on a momentous achievement. The scientific and technological feat of making the first direct detection of gravitational waves is truly monumental, and deserving of widespread recognition. . . . The IPTA is constantly improving its capabilities towards the detection of nanohertz gravitational waves, primarily from the inspiral of supermassive black hole binaries. We look forward to the day when we too have the privilege to claim a gravitational-wave detection, but today we simply raise a glass and toast the incredible success of LIGO!

And optimism has certainly not diminished. In March 2016, Stephen Taylor of NASA's Jet Propulsion Laboratory and his colleagues presented a new analysis, and predicted an 80 percent chance that detection of nanohertz gravitational waves would occur within a decade.

One thing to keep in mind is that all expectations are based on theoretical models. The strength of the gravitational wave background depends on a large number of assumptions. Models and assumptions can be wrong. Yes, galaxies contain supermassive black holes, and yes, galaxies collide and merge. But the devil may well be in the details. What is the mass distribution of supermassive black holes—in other words, how many of them are there within a certain mass range? How do galaxies and supermassive black holes evolve? How frequently do galaxies collide? If mergers occurred more often in the distant past (which is very likely), then how exactly does the merger rate decrease with time?

Other uncertainties have to do with events *after* a collision. How long does it take for the two supermassive black holes to gravitationally "sink" to the center of the merged galaxy? Will they really end up close enough to each other to emit detectable gravitational waves? This all has to do with the way in which the black holes interact with individual stars and gas clouds in the galaxy's core region—something we know hardly anything about.

There are many possible reasons our expectations about the detection of the gravitational wave background might be wrong. Maybe fewer supermassive black holes were born in the early universe. Galaxy merger rates may have been lower than what is generally assumed. It might take billions of years for supermassive black holes to get close enough to each other. There could even be millions of "stalled mergers" out there. Maybe the final spiral-in phase is much faster than theorists believe. Perhaps it's a mix of different things.

At the same time, pulsar timing array measurements—even the null results so far—are valuable pieces of the puzzle. The strength of the gravitational wave background provides astronomers with useful information about the evolution of galaxies and supermassive black holes. Thanks to those decades-long programs, theorists now have some real data to confront their pet theories with. A number of theoretical models for the merger evolution of galaxies have already been refuted, since they predicted nanohertz waves strong enough to have been detected by now. Likewise, if nanohertz waves *are* detected in the near future, their characteristics will tell us a lot about what goes on in the distant universe and at the cores of galaxy mergers.

For now, pulsar astronomers continue their meticulous work. Every two weeks or so, they check in on dozens of millisecond pulsars, adding the timing data to their ever-growing database. Slowly but surely, the sensitivity goes up, year after year. Nobody doubts that the hunt will be successful eventually. However, there will never be a LIGO-like revolutionary detection. It will be a matter of gradually increasing confidence instead.

> Fast-forward to 2030. The instruments of the past have become obsolete. Arecibo, Parkes, the Green Bank Telescope— they all ran into financial problems a decade ago, when government funding agencies decided to put their money elsewhere. The huge radio telescopes have turned into open-air museum exhibits, appreciated as monuments of cultural, industrial, and scientific heritage. The control buildings are now popular centers of science education, frequented by elementary and secondary school groups. Maintenance of the giant dishes is provided by volunteers of local astronomy clubs and ham radio organizations.
>
> In Europe, the situation is comparable, although some of the radio telescopes that took part in the original European Pulsar Timing Array project are still in use by professional astronomers. In the northeastern part of the Netherlands, the Westerbork Synthesis Radio Telescope has just celebrated its sixtieth birthday. A small onsite exhibit features the most important astronomical discoveries of the observatory, including the first convincing evidence for the existence of dark matter in galaxies back in the 1970s. The last panel of the exhibit tells the story of the detection of nanohertz Einstein waves, in the early 2020s, made possible by linking the five EPTA observatories into one "virtual" telescope almost 200 meters across. The LEAP (Large European Array for Pulsars) project, which had started a few years earlier, finally provided

the necessary jump in sensitivity to convincingly measure the gravitational wave background.

Meanwhile, pulsar astronomy has become a thriving scientific discipline. Some 20,000 pulsars have now been discovered in our Milky Way galaxy—about 10 percent of the estimated population. Among them are more than a thousand millisecond pulsars; the fastest one has an incredible spin rate of 1,130 revolutions per second. The number of known pulsar planets has grown to thirty-four, in fourteen different systems. Binary pulsars abound. One particular system grabbed everyone's attention when it was discovered in 2027, because of its small distance, extremely short period, and rapid orbital decay. The soon-to-be-launched Laser Interferometer Space Antenna is expected to pick up the faint, middle-frequency gravitational wave signal of the two orbiting bodies.

At the Jocelyn Bell International Center for Pulsar Research, scientists also study nanohertz gravitational waves on a routine basis. The International Pulsar Timing Array program now monitors about 500 millisecond pulsars. Timing accuracy has improved to something like 10 nanoseconds. Apart from a well-characterized background, five stronger individual sources of very-low-frequency waves have been discovered and localized—binary supermassive black holes in galaxies in the core regions of nearby clusters.

If there's any truth in this imaginary future scenario, it will—at least in large part—be due to a new radio observatory that's going to dwarf all others on the planet. This is not going to be a single-dish instrument like Parkes, Arecibo, or the recently completed 500-meter FAST telescope in China. Nor will it be a classic radio interferometer like Westerbork in the Netherlands or the Very Large Array in New Mexico. Instead, the Square Kilometre Array (SKA), as it is called, is a planned collection of many hundreds of radio dishes and tens of thousands of simple dipole antennas. Eventually it will have a total collecting area of one square kilometer—hence the name. The dishes and antennas will all be hooked up by fiber optics and work in unison, pouring hundreds of terabytes of raw data per second into a powerful central supercomputer. It's the largest scientific facility ever to be built by humans.

If you think Parkes in New South Wales is a small town with an unassuming center, go visit Murchison in Western Australia, on the other side of the continent. It's a loose scattering of houses, with just one combination shop / bar / gas station. A few dozen people live here, in what used to be home to the Wajarri Yamatji indigenous people. A few more live farther away, at farmsteads in the outback. All in all, Murchison Shire is about the size of Maryland and has 110 inhabitants. It's a radio astronomer's paradise.

Close to Boolardy Station, a huge cattle ranch, Australian astronomers have erected thirty-six 12-meter radio dishes, spread out over a huge area of desert. This is the Australian Square Kilometre Array Pathfinder, or ASKAP for short. Construction of the dishes was completed in 2012. Installation of the sensitive phased-array antenna feeds took another couple of years; astronomers carried out the first science observations—with just eleven dishes—in the spring of 2016.

Not far from ASKAP is another SKA pathfinder telescope, the Murchison Widefield Array (MWA). It doesn't look like a radio observatory at all. Instead, MWA consists of many dozens of antenna fields, or "tiles." Each tile contains sixteen spider-like dipole antennas, reaching no higher than some 50 centimeters. It's a technique first pioneered by the LOFAR (Low-Frequency Array) telescope in the Netherlands. ASKAP and MWA are complementary: ASKAP is one of the fastest radio telescopes in the world, designed to carry out large surveys of the universe, while MWA focuses on the lowest-frequency radio waves from the cosmos, dating back to just a few hundred million years after the big bang.

This remote outback region was chosen for its extreme radio quietness. The use of cellphones is strictly forbidden. The ASKAP control building has a metal shell to prevent leakage of radio waves from computers and electronics inside the building. One of the main sources of radio interference is high-flying planes, so radio astronomers are trying to have some air corridors relocated. The area itself is flat, hot, and dry, a huge expanse of red sand and shrubby vegetation populated by mosquitos, birds of prey, and kangaroos.

A few years from now, the Murchison Radio Observatory will be the core of the Australian part of the Square Kilometre Array. Building on the experience gained with the MWA, astronomers will construct tens of thousands of larger dipole antennas, which are shaped like Christmas trees and the height of a human being. They will be grouped together in circular stations, spread out over many hundreds of kilometers of red Australian desert. Hooked up by fiber optics and linked to a giant supercomputer in Perth, the antennas will constitute the most sensitive low-frequency radio ear that has ever been constructed.

Meanwhile, in the Great Karoo semi-arid region in South Africa, northwest of the small town of Carnarvon, two more SKA pathfinder telescopes are now in operation. HERA (Hydrogen Epoch of Reionization Array) consists of nineteen simple 14-meter wire mesh dishes. It is being expanded to contain some 350 dishes by late 2018. MeerKAT is an array of sixty-four 13.5-meter radio dishes. It will be incorporated into the first construction phase of the middle-frequency part of SKA, which is about to start.

Eventually, many hundreds of dish antennas will work in unison here to study radio galaxies and quasars, the origin and evolution of galaxies, supernova remnants, and prebiotic molecules in space. And pulsars, of course. The Square Kilometre Array (in particular the South African part) will usher in a whole new era of pulsar timing array measurements thanks to its huge sensitivity.

There's one other area of gravitational wave research in which SKA is expected to play a major role. That's the identification of the sources of the waves. "Feeling" the tiny tremors of spacetime can certainly tell you a lot about cosmic catastrophes such as stellar explosions and neutron star mergers. But scientists always want more. Come to think of it, that's a pretty natural response. If the ground beneath your feet starts to tremble, you'll look around to see what may be causing it. The more clues you can get, the better. That's why astronomers are hunting down the so-called electromagnetic counterparts of gravitational wave sources, combining as many observations as they possibly can. Using radio observatories and fast-responding optical instruments, it might be possible to actually "see" the events that have produced the Einstein waves in the first place.

Welcome to multi-messenger astronomy.

(((14)))

Follow-Up Questions

The Roque de los Muchachos Observatory on La Palma in the Canary Islands is one of the most enthralling places I've ever been. La Palma is a steep volcano, rising 2,423 meters out of the Atlantic Ocean, off the coast of Morocco. The observatory is perched on the north rim of the volcano's huge caldera. From the harbor town of Santa Cruz de la Palma, a dangerous access road with dozens of hairpin curves takes you to the rock-strewn summit, where you often look out over a layer of clouds farther down the volcano's slope. It makes you feel like you're really on top of the world, almost as close to the stars as you can get.

Late in the evening of Friday, February 28, 1997, one of the observatory's domes started to move unexpectedly. The 4.2-meter William Herschel Telescope was officially scheduled to observe a part of the sky in the constellation Serpens, the Snake. But now it moved much farther west, to a region of sky very low above the horizon. Staff astronomer John Telting snapped a couple of pictures of a small area in the northwestern part of Orion. That same night, the digital images were sent over the Internet to the University of Amsterdam in the Netherlands. Before long, graduate students Paul Groot and Titus Galama had achieved a breakthrough in the budding field of gamma-ray burst astronomy.

I know, this is supposed to be a book about gravitational waves, not gamma-ray bursts. But the two topics are very much related, as we will see later in this chapter. And for now, the story is important if we want to understand astronomers' need for quick follow-up observations of short-lived phenomena. So here's a very brief introduction to gamma-ray bursts.

In the late 1960s, mysterious bursts of high-energy gamma rays were discovered in data from the American Vela spy satellites. It took a decade before astronomers convinced themselves that the brief bursts had a cosmic origin. It took yet another decade or so before NASA's Compton Gamma Ray Observatory was launched in April 1991. One of the space observatory's goals was to collect as much data on the enigmatic cosmic explosions as possible and find out what they are. (High-energy gamma rays from space aren't observable from the ground because, fortunately, this lethal radiation is absorbed by Earth's atmosphere.)

Solving the gamma-ray burst mystery turned out to be more difficult than expected, though. Sure enough, Compton's BATSE (Burst and Transient Source Experiment) detector registered many hundreds of bursts within a few years. But it was impossible to determine their sky positions very precisely, let alone their distances. In addition, the brief bursts—sometimes they lasted for just a fraction of a second—occurred all over the place, apparently randomly. From their distribution, you couldn't possibly tell if they were relatively faint and nearby (maybe colliding asteroids or explosions on the surfaces of nearby stars) or extremely powerful events in remote galaxies.

The launch of the Italian-Dutch satellite BeppoSAX, in April 1996, changed all that. Apart from a gamma-ray monitor, the small satellite was also equipped with X-ray telescopes. The idea was that any cosmic explosion would only produce high-energy gamma rays for a very short while, but that the emission of lower-energy X-rays might last longer. Moreover, an X-ray telescope could pinpoint the burst's location in the sky much more precisely. If this information could be passed on quickly enough to astronomers on the ground, it might be possible to find a radio "afterglow," or maybe even an optical counterpart.

So when Paul Groot and Titus Galama learned that BeppoSAX had detected a burst earlier that day, they knew they had to act as fast as they could. Officially, they weren't allowed to use the information for anything other than radio observations. Moreover, the British-Dutch William Herschel Telescope, an optical instrument, was supposed to carry out other observations that night. Frustratingly, Groot and Galama couldn't reach their thesis adviser, Jan van Paradijs, for consultation. In the end, Groot decided to ignore the rules. He called up John Telting at La Palma and asked him to image the region indicated by BeppoSAX, in northwestern Orion.

Pretty soon an optical counterpart was indeed found. It became clear that the gamma-ray burst had occurred in a very distant galaxy, billions of light-years away. This also meant that the true energy output of the explosion was incredibly large—gamma-ray bursts are among the most energetic events ever observed in the universe. The revolutionary discovery led to a whole new branch of high-energy astrophysics. And it underscored the importance of quick follow-up observations of transient cosmic phenomena.

Today, such rapid follow-ups have become something of a routine in astronomy, and in many cases they are fully automated. Minutes after a gamma-ray or X-ray satellite observes a peculiar burst-like event, small robotic telescopes on the ground start to photograph the suspect region of the sky in search of a visible counterpart. Larger telescopes usually can't respond that quickly, but they will sometimes interrupt their regular observing program to help identify the culprit.

Gravitational wave signals are no exception, for very good reasons. Thus, on September 17, 2015, the European VLT Survey Telescope at Cerro Paranal in northern Chile started to scan the southern sky, searching for an optical counterpart of the gravitational wave signal that LIGO had detected three days before. As described in Chapter 11, the automated alert service was not yet operational, but LIGO and Virgo spokespersons Gabriela González and Fulvio Ricci had told the astronomers where to look, just as Paul Groot and Titus Galama told their colleague at La Palma where to look for a possible optical counterpart of a gamma-ray burst.

Together with La Palma in the Canary Islands, northern Chile is one of the world's best places to do optical astronomy. Cerro Paranal is a remote and barren mountain in the Chilean Cordillera de la Costa, some 130 kilometers south of the harbor town of Antofagasta. When I first visited the site in 1998, the only way to get there was via an 80-kilometer stretch of heavily rutted gravel road through an eerie, Mars-like landscape. The road has since been paved, but the landscape hasn't changed. The final scenes of the 2008 James Bond movie *Quantum of Solace* were shot here.

Paranal is home to one of the most productive ground-based optical observatories in the world—the Very Large Telescope (VLT). Built by the European Southern Observatory in the 1990s, the VLT consists of four identical 8.2-meter telescopes. All four of them are outfitted with a large collection of sensitive cameras and spectrographs. To support the observing program of the VLT, a smaller 2.6-meter instrument was built next to the four giants. This VLT Survey Telescope, completed in 2011, has a much wider field of view. Its huge 268-megapixel camera captures very faint stars over large swaths of sky within minutes. It's a great instrument to search for a possible optical counterpart of GW150914.

The telescope's search was fruitless, unfortunately. The same was true for other counterpart searches carried out at other observatories around the world. Maybe there was really nothing to see. After all, what kind of optical signal would anyone expect from two colliding black holes? Then again, the lack of success may have been due to something else altogether. No one knew precisely from which direction the Einstein waves had arrived. In other words, the search area covered too large a part of the sky. Nevertheless, everyone agrees on the importance of follow-up observations to find optical, infrared, ultraviolet, millimeter-wave, X-ray, gamma-ray, or radio counterparts. Any kind of electromagnetic radiation from events that produce gravitational waves might provide valuable additional information.

Aerial view of the Paranal Observatory in Chile, operated by the European Southern Observatory (ESO). The large enclosures in the center of the photo house the four 8.2-meter telescopes that make up ESO's Very Large Telescope (VLT). On the same platform is the smaller VLT Survey Telescope that was used to search for an optical counterpart of GW150914. In the background is the VISTA telescope.

So why are electromagnetic counterpart searches that essential? An analogy will make that clear. Suppose you're an ear, nose, and throat specialist in a soccer stadium. During a quiet part of the match, you hear someone sneeze. It's a weird-sounding sneeze, so since you're a real professional, you want to know everything about it. You heard that the sneeze came from somewhere on your right, but it's impossible to pinpoint it precisely by ear. Also, on the basis of the observed loudness, you only have a vague idea of the distance from which it came. You have no way to find out who sneezed—it could've been anyone.

However, if you turn your head very quickly, immediately after the sneeze, you might be able to see one person in the stadium still bent forward with her hands on her face before she reaches out for a tissue. Having localized the sneezer, you now know precisely from what distance the sound came, so you can work out the sneeze's true volume. You can also study the person's physiology, in the hope of learning more about the funny-sounding sneeze.

Two things are important here. First, if you observe something in one particular way, it's always revealing to observe the same phenomenon in a completely different way as well. If you *hear* something, you also want to *see* it. If you catch gamma rays from a cosmic explosion, you want to follow up with radio telescopes or optical instruments. If your instruments detect weak ripples in spacetime, you want to search for electromagnetic counterparts, too. Second, if the observed phenomenon is short-lived, it's essential to be quick.

For many centuries, astronomy was a slow-paced science. Planets only gradually change their positions in the sky; the constellations always look the same. A shooting star or an occasional comet might cause some excitement, but in general, astronomers never needed to hurry. What they could study one day could also be studied the next day, or the next year.

Those days are gone. Over the past decades, we have extended our horizon to billions of light-years. We've broadened our view to encompass every part of the electromagnetic spectrum. We've also greatly improved the sensitivity of our observations. As a result, we've discovered that the apparent immutability of the celestial sky is an illusion. Transient phenomena are common. In fact, the only thing that never changes is the inherent variability of it all.

Stars pulsate and vary in brightness. Red giant stars can die in supernova explosions. Dwarf stars exhibit powerful flares. If too much matter from a companion star piles up on the surface of a white dwarf, a huge thermonuclear explosion (a nova) ensues. Asteroids smash together. Comets slam into planets. Rapidly spinning neutron stars pulsate at radio or X-ray wavelengths. Black holes blast jets of particles and radiation into space. Quasars flicker. Neutron stars collide and merge. Our word *cosmos* comes from the Greek word for "order," but the universe is in constant flux and turmoil. And many transient events remain unexplained so far because of insufficient data.

By the way, in some cases the cosmos is not to blame. A bright flash in the sky that looks like a stellar explosion may actually be the glint of sunlight reflecting off the antenna of a communications satellite. Some of the bursts of gamma rays registered by NASA's Fermi space telescope are produced not in remote galaxies but by thunderstorms on Earth. And recently, scientists at the Parkes Observatory in Australia were even fooled by the microwave oven in their observatory kitchen. The Dish had registered mysterious radio signals, lasting for a quarter of a second or so. Astronomers called them "perytons," after a mythological animal. However, it was found out that perytons are generated when the door of a microwave oven is opened prematurely. No new cosmic riddle, just impatient astronomers and technicians onsite who believe that their lunch has been cooking long enough. (Here's another reminder of the importance of absolute radio silence at a radio observatory.)

Obviously, real cosmic transients are of much more interest to astronomers. And in some cases, they're still pretty mysterious. Take fast radio bursts (FRBs), for example. Like perytons, these are bursts of radio waves lasting no more than a tiny fraction of a second. And like perytons, they were first discovered by the 64-meter Parkes radio telescope. However, fast radio bursts really *do* come from outer space. They almost certainly originate in distant galaxies, just like gamma-ray bursts, although their true nature is still unknown. So far, no one has been able to respond quickly enough to the detection of a new FRB to observe it at other wavelengths. As I said before, speed is essential.

The present situation with fast radio bursts is a bit comparable to the early days of gamma-ray burst astronomy. In most cases, their distance isn't known well enough to say anything about their true energy output. And since there are no counterpart observations available at other wavelengths, it's hard to do any follow-up work. Little wonder, then, that the Dutch astronomer who helped establish the distance scale of gamma-ray bursts is eager to solve the FRB mystery, too. Between 2006 and early 2017, Paul Groot chaired the astrophysics department of Radboud University in Nijmegen. Together with colleagues in South Africa and the United Kingdom, he hopes that their MeerLICHT project will yield the breakthrough.

With MeerLICHT, the response time for counterpart searches is essentially brought to zero. MeerLICHT is a relatively small, 65-centimeter robotic telescope recently installed at the Sutherland Observatory in South Africa. It is programmed to *always* look in exactly the same direction as MeerKAT, one of the South African pathfinder observatories for the Square Kilometre Array, some 250 kilometers farther north. If the radio telescope happens to observe an FRB (or some other transient source) with an optical counterpart bright enough to be visible, the image will automatically be captured by the robot telescope. In terms of speed, you can't beat simultaneity.

You might think this would also be a promising strategy to discover optical counterparts of gravitational waves. However, with Einstein wave detectors such as LIGO and Virgo, it's hard to have an optical telescope always look in the same direction. The simple reason being that LIGO and Virgo are omni-sensitive—they will register strong-enough gravitational waves no matter from which direction they arrive here on Earth. And of course, you can't have sensitive optical telescopes cover the whole sky continuously. The field of view of a telescope is usually much smaller than the apparent size of the full Moon, so astronomers have to live with the fact that they can't look everywhere at once.

The obvious solution is the alert system that has been designed for LIGO and Virgo. As soon as a plausible gravitational wave candidate has been detected, astronomers are informed about its direction of origin so they can put telescopes and space observatories to work. In principle, this can all be automated. The data streams of the laser interferometers are continuously checked by detection algorithms. When a signal looks significant enough to warrant further analysis— as in the case of GW150914 and GW151226—a rough sky position can be calculated from the data. This is then disseminated over the Internet to all observers who have a formal agreement with the LIGO-Virgo Collaboration. If they operate robot telescopes, the first images of a possible counterpart can be achieved within minutes after the actual Einstein wave detection.

Astronomers have thought a lot about what kind of counterparts they can expect with a gravitational wave, and for how long these might be visible. To answer those questions, they first need to know what kind of cosmic events produce observable gravitational waves.

The existing laser interferometers are sensitive to gravitational waves with a frequency between, say, 10 and 1,000 hertz. Those are mainly produced by collisions and mergers of neutron stars and black holes. Such events are "visible" to LIGO and Virgo over large distances. Eventually, when the advanced detectors reach their full design sensitivity, astronomers will be able to observe neutron star mergers out to a few hundred million light-years. For the collision between a neutron star and a black hole, the corresponding distance is well over a billion light-years, thanks to the larger mass of the black hole. The coalescence of two black holes can even be observed out to a few billion light-years, provided the holes are massive enough.

So what would you expect to see with an optical telescope, or at infrared, X-ray, or radio wavelengths? Well, that depends. A "clean" black hole merger would not produce any electromagnetic radiation. After all, such events are just "storms in the fabric of spacetime," as Kip Thorne describes them. There's no stuff around—no atoms, no molecules, no anything—that might give off any type of radiation. The only way a black hole merger can communicate with the rest of the universe is through gravitational waves.

That's why the counterpart hunters were a bit disappointed by the fact that GW150914 was produced by two coalescing black holes. Conceivably, the site of the cosmic collision could contain some matter in the form of interstellar gas and dust, but given the tremendous gravitational pull of the two black holes, it would probably not be much. And without matter to heat up or to create shock waves in, chances were slim that the event would have produced any detectable electromagnetic radiation. (Which didn't stop astronomers from carrying out counterpart searches anyway.)

But in the case of a neutron star merger or a collision between a neutron star and a black hole, the story is very different. A neutron star contains at least 1.4 solar masses' worth of ordinary nuclear particles. Of course, if two neutron stars collide, the end result will probably be a black hole. And if a neutron star crashes into a black hole, most of its mass will disappear for good. But in both cases, a significant quantity of matter may be heated to extremely high temperatures and get ejected into space at a considerable fraction of the speed of light. When this blast wave runs into the surrounding interstellar material, however insubstantial that may be, powerful shock waves will produce electromagnetic radiation at a wide range of wavelengths. All in all, collisions involving at least one neutron star are expected to produce spectacular cosmic fireworks.

In fact, this is where the link between gravitational waves and gamma-ray bursts comes into play. Already back in the early 1990s, some astrophysicists argued that gamma-ray bursts might be produced by neutron star mergers in distant galaxies. That was well before the distance scale of the explosive events had been established. Today, almost no one doubts that neutron star mergers are responsible for at least a large fraction of the observed gamma-ray bursts.

What's important to know here is that gamma-ray bursts can be divided into two groups, each of which represents a different population of cosmic phenomena. Short gamma-ray bursts last for just a fraction of a second, while long gamma-ray bursts have a duration of several seconds to a couple of minutes. The long bursts are probably extremely powerful supernova explosions. Also called hypernovae, they may result from very massive, rapidly rotating stars that catastrophically collapse into black holes at the end of their brief lives. For the short bursts, various scenarios have been proposed, but the neutron star merger model is by far the most popular.

So let's concentrate on the short gamma-ray bursts. For some of them, faint X-ray and optical afterglows have been detected. These afterglows last much longer than the original burst of gamma rays— even longer than a day. Naively, you might now think that we know precisely what to expect from counterpart searches of gravitational waves. After all, we may be talking about exactly the same physical phenomenon—neutron star mergers. If a burst of gravitational waves is also produced by a neutron star merger, wouldn't you expect an almost simultaneous flash of high-energy gamma rays, sometimes followed by a faint afterglow?

Unfortunately, it's not that straightforward. The reason is that gamma-ray bursts are highly "beamed." Their incredible amount of instantaneous energy is preferentially emitted in two opposite directions. For hypernovae (long bursts), the beaming occurs along the rotational axis of the collapsing star. For neutron star collisions (short bursts), it may occur perpendicular to the orbital plane of the merging stars. Apparently this is the direction in which most of the matter is ejected out of the system at unimaginably high velocities very close to the speed of light.

If we happen to look into one of those two beams (or jets), we observe the titanic explosion as a gamma-ray burst. But if we happen to look at it from the side, we don't see a gamma-ray burst at all, and not much of an afterglow, either. In other words, many neutron star mergers will not be observed as gamma-ray bursts, and the true number of neutron star mergers in the universe is much larger than the number of short gamma-ray bursts that astronomers detect.

In contrast, gravitational waves are emitted in all directions (although not necessarily at exactly the same strength). Even if a neutron star merger is not observable as a short gamma-ray burst, because of its orientation in space, it may still be observable as a source of Einstein waves. There's one caveat, though: such waves are weak and hard to detect. As a result, we can only expect to observe them from neutron star mergers within, say, a few hundred million light-years.

So yes, there may be a link between sources of gravitational waves and short gamma-ray bursts, but it's complicated. In fact, it's a bit like the relation between neutron stars and pulsars. Rapidly spinning, highly magnetized neutron stars produce rotating "lighthouse" beams of radio waves, as we saw in Chapter 6. If the orientation is favorable, we can detect those neutron stars as pulsars, even at distances of tens of thousands of light-years. The true number of neutron stars is of course much larger than the number of pulsars we observe. However, the isotropic emission of neutron stars—the radiation that is emitted in all directions—is very weak. That's why a neutron star that is *not* observable as a pulsar can be seen only when it is pretty close—a few hundred light-years or so.

Now we're getting somewhere. LIGO and Virgo will be able to detect the gravitational waves of a pair of merging neutron stars only if the catastrophe takes place within a few hundred million light-years. If such a collision produces strongly beamed radiation, there are two possibilities: either one of the beams is pointed in our direction (small chance) or both beams miss the Earth (much larger chance). In the first case, we expect to see a tremendously bright short gamma-ray burst and a pretty conspicuous afterglow at many different wavelengths. Such an event will certainly be registered by orbiting gamma-ray observatories. In the second case, however, we need to know what kind of isotropic emission the merger event is likely to produce.

Theorists think they know the answer to that question. Immediately after the collision, the matter that's blasted into space is extremely hot. Moreover, it's not as unbelievably dense as it was in the neutron star phase. Suddenly, nuclear reactions can take off again. And they will. Disintegrating clumps of closely packed neutrons fly around. Individual neutrons decay into protons—positively charged nuclear particles. Together, protons and neutrons form massive lumps of nuclear matter that immediately start to break up into smaller, more stable nuclei. Radioactive elements rapidly decay away, producing large amounts of radiation, mostly at red and infrared wavelengths. What's left is an expanding and slowly cooling cloud of heavy elements, including such precious ones as gold and platinum.

According to calculations by Edo Berger of the Harvard-Smithsonian Center for Astrophysics in Cambridge, Massachusetts, the collision of two neutron stars may produce no less than ten times the mass of the Moon in pure gold. In fact, almost the whole cosmic inventory of this precious metal—including the gold in your wedding ring, bracelet, or watch—has probably been created in neutron star collisions.

The estimated amount of energy that's emitted in the post-collision nuclear cauldron is less than the energy of a traditional supernova explosion. However, it's about a thousand times *more* than what is produced during a regular nova (a thermonuclear explosion on the surface of a white dwarf star). That's why this event is often called a "kilonova." For obvious reasons, another popular name is "bling nova."

In the summer of 2013, Nial Tanvir of the University of Leicester in the United Kingdom and his colleagues were the first to observe the expected kilonova emission from a short gamma-ray burst. The burst had been detected on June 3 in a galaxy at a distance of almost 4 billion light-years. Using the Hubble Space Telescope, Tanvir's team observed the fading fireball on June 12. The discovery is generally seen as firm proof that short gamma-ray bursts are indeed the result of neutron star mergers. And since the kilonova radiation is emitted in all directions, this is also the kind of electromagnetic counterpart that you might expect from a neutron star merger that is *not* visible as a gamma-ray burst.

So now we know what to expect in the aftermath of a gravitational wave detection. If the spacetime ripples are produced by merging black holes, there probably won't be any kind of electromagnetic radiation at all. But if at least one neutron star is involved in the collision, you may expect a brief initial burst of high-energy, bluish light, followed by a slowly fading glow at red and infrared wavelengths. At a later stage, the expanding material may also start to emit radio waves. Of course, this is just the current theoretical wisdom; the universe may have many surprises in store for us.

One important additional piece of information that the discovery of a counterpart will bring is the distance to the source of the gravitational waves. For both GW150914 and GW151226, the existing distance estimates are pretty uncertain. They're purely based on the observed amplitude of the waves and on theoretical models. But if a counterpart is discovered in a remote galaxy, it's easy to determine the galaxy's distance. All you have to do is measure the galaxy's redshift, as explained in Chapter 9. And if you know the distance, you can deduce the energetics of the collision, including the energetics of the gravitational waves. That would be a nice way to test and improve the existing models.

All in all, hunting for electromagnetic counterparts and carrying out follow-up observations seem to be the right things to do. There's a lot of interest from many astronomical fields. Dozens of teams have applied for a collaborative agreement with LIGO and Virgo to receive alerts as soon as a new gravitational wave signal is detected. Together, those searches cover the whole electromagnetic spectrum, from the longest radio wavelengths to the shortest gamma rays. They also use a wide variety of instruments, from small automated cameras to the largest optical telescopes and radio dishes, as well as Earth-orbiting satellites. Rest assured that the world will be watching as soon as the interferometer mirrors start wiggling again.

I've now explained why rapid follow-up observations of gravitational wave signals are important, and what kind of counterparts we might expect. There's one major problem, though. As mentioned before, the search areas are much too large. At least, that was the case during the first observing run of Advanced LIGO. The only way to deduce the direction of origin of a gravitational wave is by precisely timing its arrival at multiple detectors. But if you have only two detectors at your disposal, it's generally not possible to find the answer. It's easy to see why.

The two LIGO detectors, at Livingston and Hanford, are separated by some 3,000 kilometers. Imagine a straight line connecting the two observatories and extending into space in two directions. Now suppose the cosmic collision that produced the gravitational waves occurred exactly on that line. In that case, the waves would take 0.01 seconds to travel from the first detector to the second (remember that gravitational waves propagate at the speed of light, which is 300,000 kilometers per second). So if Hanford sees a signal 0.01 seconds earlier than Livingston, you'd know that the event occurred on the Hanford side of this connecting line. If the Hanford detection occurs 0.01 seconds later, the waves came in from the opposite side.

Of course, there's only a minute chance of such a precise lineup. In most cases, the time difference will be less than 0.01 seconds, since the waves will arrive under a certain angle with respect to the line that connects the two observatories. (If they arrive perpendicular to the connecting line, there won't even be a time difference at all—the two detectors will register the signal simultaneously.) But in that case, you wouldn't know from which direction the waves came. The only thing you can do is draw a circle on the sky and conclude that the collision must have happened somewhere on that circle. The shorter the time difference is, the larger this circle will be.

Some other characteristics of the detection may tell you why one part of the circle is more likely to contain the source than some other part. Still, you're usually left with a giant banana-shaped segment of the sky where the merger could have happened. Quickly finding a counterpart thus requires fast coverage of a huge part of the celestial sphere. But such a large swath of sky will probably contain many dozens of suspect objects: little pinpricks of light that weren't there a month before and will be fading away in the days to come. For each of them, you'd need to make sure it's not some other type of transient, such as a distant supernova, a stellar flare, or whatever. In the end, you'll probably never know for sure if you've really identified the source of the observed Einstein waves.

With the official dedication of Advanced Virgo on February 20, 2017, things look much better, of course. When three detectors observe the same gravitational wave signal, there are also three pairs of them: Livingston-Hanford, Livingston-Virgo, and Hanford-Virgo. Three pairs mean three different ways to do the same analysis, so you'll end up with three circles (or banana-shaped segments) on the sky. These will overlap in one relatively small region, so that's where you should hunt for counterparts. In fact, I wouldn't be too surprised if the first Einstein wave counterpart has already been found and studied by the time this book hits the market, although at the time of writing, Advanced Virgo is still encountering problems with the fused silica suspension wires of the mirrors. (Meanwhile, Advanced LIGO's second observing run started on November 30, 2016; as of April 2017, six "event candidates" had been identified.)

A few years from now, a fourth laser interferometer will be operational in Japan. In the future, number five goes online in India (more on those two in Chapter 16). As you can imagine, more detectors yield an even more precise localization. With all these rapid developments, it can only be expected that follow-up studies of gravitational wave sources will soon grow into a mature and promising field of astrophysics.

Electromagnetic counterpart searches will not only be carried out by optical and radio telescopes on the ground. In fact, the first instrument to claim success might well be a space observatory. After all, the highest-energy electromagnetic radiation—gamma rays and X-rays—cannot be observed from the ground at all. Several teams of gamma- and X-ray astronomers also have agreements with the LIGO-Virgo Collaboration. They're ready to point their space telescopes in any direction that the interferometers come up with.

For instance, NASA's Swift satellite, launched in November 2004, has been involved in the search for quite some time already. Swift was designed to detect and study gamma-ray bursts. It is equipped with a gamma-ray detector, an X-ray telescope, and an ultraviolet/optical telescope. All on its own, Swift can detect new gamma-ray bursts, determine their sky positions, and search for optical counterparts. But according to principal investigator Neil Gehrels of NASA's Goddard Space Flight Center in Greenbelt, Maryland, the successful mission is also capable of quickly looking for X-ray, UV, or optical counterparts of gravitational wave sources. In the past, Swift has carried out follow-up observations for a number of LIGO/Virgo triggers. It even spent a couple of days on the Big Dog event, the infamous blind injection in September 2010 that was described in Chapter 11.

Another NASA instrument, the Fermi Gamma-Ray Space Telescope, could also play a decisive role, according to Gehrels. Fermi was launched in June 2008. Its wide-angle gamma-ray detectors cover about half the sky. So if a gravitational wave signal is accompanied by a burst of high-energy gamma rays, there's about a 50 percent chance that Fermi will see it. In that case, Swift may follow up on the Fermi detection to determine a more precise position. Within minutes, ground-based optical telescopes could start searching a much smaller area than can be provided by LIGO and Virgo alone. (Too bad Neil Gehrels won't be around to witness the results of such a search—he died in early 2017, at age sixty-four.)

What about instruments on the ground? Well, some large telescopes are equipped with wide-field cameras to repeatedly map the heavens. We already encountered the VLT Survey Telescope at Paranal, with its 268-megapixel camera. Another is the 520-megapixel Dark Energy Camera on the 4-meter Blanco Telescope at the Cerro Tololo Inter-American Observatory in Chile. And then there are the two 1.4-gigapixel Pan-STARRS cameras, mounted on 1.8-meter telescopes at the Haleakala Observatory on the island of Maui, Hawaii. However, these large instruments are not really designed to do quick follow-up studies of transient objects. Smaller ones are in a much better position for that. One of those smaller instruments operates from a famous site in southern California—the Palomar Observatory.

At Palomar Mountain, northeast of San Diego, the relatively small Samuel Oschin Telescope is easy to miss. Tourists who drive up to the observatory will generally marvel at the huge dome of the 5.1-meter Hale Telescope, take a look at the giant reflector from the visitors' gallery, buy a souvenir at the gift shop, and return to their car. Fair enough—the Hale Telescope (named after astrophysicist George Ellery Hale) is a truly impressive instrument. It debuted in 1948 and served as the world's largest telescope for more than twenty-five years. When I started out as a teenage amateur astronomer in the early 1970s, the Hale Telescope was what the Hubble Space Telescope is to a younger generation. Walking beneath and around this majestic instrument is awe-inspiring.

The much smaller Samuel Oschin Telescope is a short drive from the dome of the larger scope. Its primary mirror is only 1.2 meters across. Also known as the Palomar Schmidt (for its optical design), it has an enormous field of view, more than twelve times the width of the full Moon. The telescope was used in the 1950s to produce the famous Palomar Observatory Sky Survey—a huge photographic atlas of the northern sky.

Today, Palomar astronomers such as Edwin Hubble (after whom the space telescope was named) would hardly recognize the instrument. A huge shutter the size of a ping-pong table is mounted on top of the telescope. The tube has been cut open to accommodate a large, deep-cooled CCD camera and additional optics. Cables and electronics are all over the place. Moreover, it's now fully automated—there's no one around at night, when the telescope scans the skies. Welcome to the Zwicky Transient Facility (ZTF), one of the fastest sky mappers on the planet.

According to ZTF project scientist Eric Bellm of Caltech in Pasadena, the instrument can take 30-second exposures every one and a half minutes or so. Because of the sensitive electronics, each image reveals almost the same number of stars as the photographic glass plates of the 1950s, which were exposed for about an hour. In principle, ZTF could photograph the whole visible sky in one single night, producing a stunning data flow of some 100 megabits per second.

The Zwicky Transient Facility is named after Caltech astronomer Fritz Zwicky, who carried out some of the first astronomical surveys looking for supernova explosions in other galaxies. ZTF is also on the lookout for distant supernovae and other short-lived phenomena. But, of course, the instrument can respond to gravitational wave alerts as well. Within a minute after receiving an alert, the telescope and its dome can be slewed in the right direction to start hunting for optical counterparts.

In the Southern Hemisphere, one of the future competitors of the Zwicky Transient Facility will be the BlackGEM project in Chile. BlackGEM will start out in 2018 as a small array of three automated 65-centimeter telescopes. If more funding can be secured, the array may expand to five or even fifteen identical telescopes, each equipped with a sensitive CCD camera. BlackGEM's principal investigator is Dutch astronomer Paul Groot, who also found the first optical counterpart of a gamma-ray burst. In fact, Groot's MeerLICHT telescope, described earlier in this chapter, is a prototype of the BlackGEM instruments.

BlackGEM has a couple of advantages over other counterpart search projects. First, it is dedicated to follow-up studies of gravitational wave detections—that's the main science goal. (In contrast, the Zwicky Transient Facility can follow up on at most a handful of triggers per month, given its other commitments.) Second, because the array contains multiple telescopes, BlackGEM is quite flexible. If the search area provided by LIGO and Virgo is very elongated, as it was for GW150914 and GW151226, each telescope can focus on just one part of the banana. If the search area is small, or if a counterpart has been detected, the telescopes can observe in concert, resulting in a much higher sensitivity.

Cerro La Silla in Chile, northeast of the harbor town of La Serena, has been selected as the site for BlackGEM. The atmosphere above La Silla is much steadier than that above Palomar Mountain, so the observing conditions are much better—that's a third advantage. In the 1960s, La Silla became home to some of the first telescopes of the European Southern Observatory (ESO). Today, most of ESO's activity has moved much farther north, to Cerro Paranal, but there's still a lot going on at La Silla. Perched on a saddle-shaped ridge, the observatory

The BlackGEM telescope array will consist of a number of robotic 65-centimeter telescopes. As soon as laser interferometers such as LIGO and Virgo detect a new gravitational wave signal, BlackGEM will scan the skies for possible optical counterparts.

is surrounded by gently rolling hills that stretch out toward the distant horizon. It's a very tranquil place, frequented by wild donkeys and Atacama foxes. On a clear day (and there are many), you can easily see the domes of the Carnegie Institution for Science's Las Campanas Observatory, some 25 kilometers away.

———

So many remote mountaintops, so many observatories, so many sensitive instruments and dedicated astronomers who want to solve the mysteries of the universe. Will the Zwicky Transient Facility at Palomar Mountain or BlackGEM at Cerro La Silla be the first to find an optical counterpart of a gravitational wave detection? Or will follow-up studies of the cosmic collisions depend on radio observations or X-ray and gamma-ray measurements? Radio astronomers working on the Square Kilometre Array and its various pathfinder telescopes are already discussing the best response strategy. X-ray astronomers hope to realize an all-sky monitor that could be mounted on the International Space Station. Existing observatories, both on the ground and in space, all try to get a piece of the pie. New projects are coming online on a regular basis. Success is around the corner. It's just a matter of time.

Eventually, even the dream of looking everywhere at once may come true. The future 8.4-meter Large Synoptic Survey Telescope (LSST) will image the visible sky three times per week at an incredible level of sensitivity, using a 3-gigapixel camera. LSST is under construction at Cerro Pachón, yet another observatory peak in northern Chile. It will discover tens of thousands of short-lived transients, such as supernovae, flare stars, and asteroids, and map the positions and shapes of billions of galaxies, to study the structure and evolution of the universe at large.

Meanwhile, at Cerro Tololo, just north of Cerro Pachón, the futuristic Evryscope is already snapping a quarter of the entire sky every two minutes. Led by Nicholas Law of the University of North Carolina in Chapel Hill, it's a constellation of twenty-seven automated amateur-size telescopes that act together as a giant astronomical fisheye lens. Obviously, because of its small optical aperture, the Evryscope doesn't go as deep as the LSST does—it won't see very faint stars or extremely fine detail. But put more Evryscopes around the globe, and you obtain continuous all-sky surveillance of the cosmos.

This is the future of astronomy. *Every* conceivable type of observation, *every* part of the sky, *all* the time. Photons of every possible wavelength, from the highest-energy gamma-rays to the lowest-frequency radio waves. Subatomic particles from space, such as cosmic rays and neutrinos. Minute ripples in the very fabric of spacetime. Together, all those messages constitute a treasure trove of information about the wonderful universe we inhabit.

LIGO's revolutionary detection of gravitational waves truly marked the birth of multi-messenger astronomy.

(((15)))

Space Invaders

The food was delicious, but the reception was terrible.

I'm referring to the cellphone reception. As part of an international group of journalists, I was invited by the European Space Agency (ESA) to attend the launch of their LISA Pathfinder spacecraft from the Guiana Space Center in Kourou, French Guiana. The day before the launch, we enjoyed a great lunch at the restaurant Carbet des Maripas, on the bank of the Kourou River, in the middle of the jungle. Some of us even took a short trip in one of the colorful pirogues. It all felt like a great vacation; there seemed to be no need to be in touch with the rest of the world.

Until Gaele Winters, ESA's director of launchers, made an announcement somewhere between the second and third courses. Because of a technical problem with a temperature sensor in the upper stage of the rocket, the launch would not proceed as planned, in the very early morning of December 2, 2015. Of course, all the assembled journalists wanted to call editors, update blogs, post Facebook messages, or send out tweets. But Carbet des Maripas didn't have Wi-Fi, and there was no cellphone signal, either.

Luckily, someone found out that there was intermittent coverage down by the riverside. Not more than one bar, but sufficient for most purposes. So there we were, huddled together on the small wooden pirogue jetty, with our smartphones held up high in the air. It must have been a funny sight.

The sensor problem was solved quickly, and the launch was delayed by just one day, to 1:04 AM local time on December 3. Perched on its small European Vega launcher, the spacecraft thundered into the night sky, its glowing exhaust playing hide-and-seek with a layer of clouds. Within minutes it was gone. It was a picture-perfect launch, with flames, smoke, roaring noise, and all. In the control room, people cheered and hugged. Some of them had been working on the project for more than fifteen years. Champagne flowed. Some tears, too.

Just three months earlier, I had met LISA Pathfinder up close and personal in the clean room of the Industrieanlagen-Betriebsgesellschaft in Ottobrunn, Germany (just south of Munich), where it had been tested. The address was Einsteinstrasse 20—very appropriate for a mission that would test technologies to detect gravitational waves in space. The spacecraft, about the size of a hot tub, was all wrapped up in gold-colored thermal blankets and mounted on its propulsion module. Elsewhere in the clean room, the giant crate in which it would be transported to French Guiana was already waiting.

I knew that deep inside LISA Pathfinder were two massive, highly polished gold-and-platinum cubes the size of small paperweights, meant to be in undisturbed free fall a few weeks after launch. The technological heart of the spacecraft also contained a miniature interferometer, with lasers, mirrors, and photo detectors. I found it hard to imagine that the delicate equipment could ever survive the truck ride to the United Kingdom (for final preparations), the flight to Kourou on an Antonov freight plane, the violent launch into space, and the subsequent journey to its operational location in solar orbit.

"This is our first step to observe gravitational waves from space," said project scientist Paul McNamara of ESA's European Space Research and Technology Centre (ESTEC) in Noordwijk, the Netherlands. "LISA Pathfinder is opening the door to the future." Three months later, at the launch, ESA director of science Alvaro Giménez had a few more quotes up his sleeve: "blaze new paths," "uncharted territory," "a new chapter in science," and, the one I liked best, "I believe Einstein would be pleased." Einstein would be flabbergasted, I'd reckon.

———

So what's LISA Pathfinder all about? Well, the name says it all. It's a pathfinder mission for the Laser Interferometer Space Antenna—a future gravitational wave antenna in space. LISA is going to be a giant space version of LIGO. Using mirrors and telescopes, it will bounce laser beams between three spacecraft flying in formation a few million kilometers apart. Sensitive interferometers will measure tiny changes in the separation between cubic "test masses," located inside the three craft, caused by passing low-frequency gravitational waves.

No one has any experience with detecting Einstein waves in space. Building and launching LISA without testing out the necessary technologies in advance would be a giant leap, like asking Orville and Wilbur Wright to forget about their Wright Flyer and start building a Boeing 747 instead. In a sense, LISA Pathfinder is the Wright Flyer of space-based gravitational wave astronomy.

In a ground-based detector, the mirrors on both ends of the interferometer arms serve as test masses. They move slightly toward and away from each other as a gravitational wave passes by. As we have seen, the changes in their separation are extremely tiny—much smaller than the diameter of a proton. That's why the mirrors need to be isolated from every possible high-frequency vibration that might be a product of their surroundings. In fact, that's the main challenge for laser interferometers such as LIGO and Virgo.

In space, there are no passing trucks or slamming doors. It's a much quieter environment. But even so, numerous unwanted forces are at play. Satellites are buffeted by solar radiation—sunlight exerts a small but measurable pressure on any surface it hits. Micrometeoroids fly in at irregular intervals, from all directions. The same is true for the smattering of gas particles that have evaporated away from the atmosphere of the Earth and the other planets. Charged particles blown into space by the Sun, small temperature changes, magnetic fields, high-energy cosmic ray particles—there's a whole range of disturbing factors that could ruin your Einstein wave measurements.

The best way to shield your test mass from all those effects is to encapsulate it in a hollow spacecraft. Solar radiation pressure or the impact of a dust particle can cause the spacecraft to veer off course. But if that happens, it could use thrusters on the outside to correct its position relative to the test mass inside. The test mass itself would then be influenced only by the gravity of the Sun and the planets. In fact, that's what is meant by "free fall."

However, it's not that easy—nothing ever is. Even within the hollow spacecraft, tiny forces act upon the test mass. There will always be some gas atoms flying around, even in the highest-quality vacuum. Temperature changes still play a role, as do residual magnetic fields. A slow buildup of electrical charge on the test mass can cause a tiny drift. The small gravitational forces from the spacecraft itself are never completely symmetric. Moreover, those gravitational forces will change over time because the thrusters use up fuel. If you want to know how well you've succeeded in creating a truly free-falling test mass, you need to measure all those tiny forces and accelerations. But of course, that's impossible if the encapsulating spacecraft is always "following" the test mass. There's just no measurement reference.

That's where the *second* test mass comes in. The residual effects we're after will never be precisely the same for the two test masses. If they're both in perfect, undisturbed free fall, their mutual distance and orientation would remain constant. Instead, minute forces are at play within the hollow spacecraft. As a result, the two test masses will slowly start to drift with respect to each other. If you can measure that tiny drift, you have a good indication of the level of "quietness" you have achieved.

The main goal of LISA Pathfinder is to demonstrate the ability to create a really quiet, undisturbed environment. The cubic test masses used in the spacecraft are 46 millimeters on a side. They're composed of an alloy of 73 percent gold and 27 percent platinum. The material was chosen for its low magnetic susceptibility and high density: each cube weighs almost 2 kilograms. Similar test masses will be used in the future in the full-scale LISA detector. At a cost of some $70,000 each (not counting the much more expensive precision machining of the cubes), the LISA Pathfinder test masses may well be the most expensive solid pieces of metal ever flown into space. They're certainly the geekiest paperweights you can imagine.

The two gold-and-platinum cubes will be floating in individual small cavities deep within the heart of the spacecraft. Those molybdenum enclosures are 38 centimeters apart. They, too, are cubic in shape, measuring 54 millimeters on a side. When a test mass is centered within its cavity, there's only about 4 millimeters of space left on all six sides. Those are really tight prisons. And, of course, the cubes are not supposed to ever touch the inner walls of the enclosures.

Here's how LISA Pathfinder's engineers achieve this remarkable feat. The six walls of the first enclosure are outfitted with capacitive sensors. These electrodes precisely measure the separation between each wall and the free-floating cube, which we will call test mass 1. As soon as the cube is no longer perfectly centered (most likely due to solar radiation pressure on the outside of the spacecraft or some other external force), microthrusters correct the spacecraft's position by puffing out tiny amounts of nitrogen gas. As a result, the whole spacecraft is "following" test mass 1 in its orbit around the Sun. So far, so good.

However, test mass 1 is not in perfect free fall. As mentioned above, all kinds of tiny effects are likely at play. Scientists want to quantify those effects precisely, to learn how successful they are in creating an undisturbed environment. We already saw that the residual forces act slightly differently on the two cubes. Over time, test mass 1 and test mass 2 will slowly start to drift with respect to each other. Since the spacecraft is following test mass 1, it won't be long before test mass 2 will be hitting the inside of its own enclosure.

Now here's the trick: the electrodes on the inner walls of the second enclosure are used to actively force test mass 2 back into position as soon as it starts to drift. The tiny electrical currents needed to achieve

This cutaway drawing of the innards of the European Space Agency's LISA Pathfinder spacecraft shows the two gold-and-platinum cubic test masses in their molybdenum enclosures. In between the two test masses is LISA Pathfinder's small interferometer.

this are indicative of the remaining relative motions and accelerations between the two cubes. The smaller the necessary corrective forces, the better.

LISA Pathfinder is also equipped with a small interferometer. It consists of two laser beams and twenty-two mirrors and beam splitters. The interferometer is located in between the two test mass enclosures. It accurately monitors the tiniest changes in distance and orientation between the two gold-and-platinum cubes. This is to demonstrate the possibility of measuring separations in space at picometer levels—the interferometer measurements are thousands of times more sensitive than what the capacitive sensors can achieve.

In fact, LISA Pathfinder is testing almost every new technology that the future Laser Interferometer Space Antenna will apply. The only thing Pathfinder will *not* do is actually measure gravitational waves— it's too small for that.

But why would anyone want to measure gravitational waves in space? Well, as you will remember, ground-based detectors such as LIGO and Virgo are sensitive to Einstein waves with frequencies between 10 and 1,000 hertz or so. On Earth, you can't go much lower: environmental "seismic noise" is just too strong below a few hertz. In the quiet environment of space, however, it's perfectly possible to measure those lower-frequency waves, provided the arms of your interferometer are long enough. LISA will have arm lengths of a few million kilometers. That means it will be sensitive to gravitational waves between 1/10,000 hertz (100 microhertz) and 1 hertz. In fact, LISA will nicely fill the gap between the high-frequency measurements of ground-based interferometers and the nanohertz observations from pulsar timing arrays, described in Chapter 13.

So do astronomers expect to detect spacetime ripples in this middle-frequency band? Yes, they certainly do. Tight white dwarf binaries in our galaxy continuously produce gravitational waves in this range. The same is true for binary stellar-mass black holes a few months to years before they collide and merge. Moreover, the space observatory will be able to observe the coalescence of supermassive black hole binaries in galaxies all over the universe. I'll get back to those candidate sources at the end of the chapter, but rest assured that astronomers have always been convinced of the need to go into space.

But being convinced of the need is not enough to get an ambitious and expensive space mission going. The LISA project has a long and convoluted past, and it has met numerous hurdles and setbacks over the decades, as the following historical account will show.

The very first ideas for a space-based Einstein wave detector date back to the mid-1970s. Back then, LIGO was still a distant dream, but Rai Weiss had already worked out most of the details in his 1972 publication in MIT's *Quarterly Progress Report*. Originally, he envisioned a kilometer-sized laser interferometer on the ground. But wouldn't it be much better to build the thing in space, where you wouldn't have to worry so much about external vibrations and mirror suspension?

Weiss discussed the concept with Peter Bender of the University of Colorado in Boulder during a 1974 dinner meeting. Ever since, Bender has been involved in turning the space dream into a reality—he is generally seen as one of the founding fathers of LISA. By the way, the original plan wasn't called LISA; it went by the name SAGA, for Space Antenna for Gravitational-Wave Astronomy. It took over a decade before the SAGA idea evolved into a serious mission concept, called LAGOS (Laser Antenna for Gravitational-Wave Observations in Space). By that time, LIGO was in its early development phase.

The LAGOS concept featured three separate spacecraft that together would form a huge V-shape with legs of 1 million kilometers, trailing the Earth in its orbit around the Sun. Laser beams would be bounced between the "mother spacecraft," at the vertex of the V, and free-floating test masses in the two "daughter spacecraft." Through interferometry of the reflected laser light, tiny changes in the arm length could be detected. It would be like building a giant LIGO in space. (It wouldn't matter too much that the arms of the interferometer were not at right angles to each other but at 60 degrees.) Thanks to the orbital motion around the Sun, the sky position of any continuous source of gravitational waves could be precisely triangulated over the course of a year. All in all, LAGOS was a very powerful and ambitious concept. However, potential funding agencies felt the plan was too premature, too risky, and too expensive—which it probably was.

But if scientists are convinced of the value of an idea, they won't give up easily. A new proposal was put to ESA by German physicist Karsten Danzmann in 1993, the year in which he moved from the Max Planck Institute for Quantum Optics in Munich to Hannover to establish a new gravitational wave group there. ESA was accepting proposals for the third medium-class mission (M3) in its Horizon 2000+ space science program, and Danzmann thought this would be a good fit for that program. And the timing was certainly right. One year earlier, in 1992, the National Science Foundation had signed a cooperative agreement with MIT and Caltech to build LIGO. Hanford and Livingston had already been chosen as the two sites for the future ground-based interferometer. In Italy, the Virgo project had just been approved.

The Laser Interferometer Space Antenna, as the new concept was called, was even more ambitious than LAGOS. Danzmann's proposal called for *six* spacecraft—two at each vertex. Moreover, the triangle would measure 5 million kilometers on a side. The spacecraft would be outfitted with lasers, beam splitters, telescopes, mirrors, and photo detectors. From every vertex of the triangle, coherent laser beams would be sent off to the other two corners. So, in fact, LISA would consist of three huge, overlapping interferometers. With the three interferometers working in unison, the instrument would be able to measure the polarization of gravitational waves—the fact that the wave's amplitude is not the same in every direction. This would provide additional information about the orbiting bodies, including their orientation in space and their spin rates.

LISA was not selected as the Horizon 2000+ M3 mission—it was much too ambitious. But Danzmann and his colleagues reproposed it as a potential "cornerstone mission" in the European space science program. Pretty soon, however, it became clear that the project would be too expensive for ESA alone, so they sought to collaborate with NASA on a joint program. Each agency would pay about half of the total cost, estimated to be somewhere between $1.5 billion and $2 billion. In 1996, the first international biennial LISA Symposium was organized, taking place in the United Kingdom. Two years later, the scientists proposed a technology demonstrator, called ELITE (European LISA Technology Experiment). Momentum was growing, albeit much more slowly than many had hoped for.

It took until 2010 before a detailed mission plan was finally completed. By then the original ELITE proposal had evolved into LISA Pathfinder. However, work on the demonstrator mission had already experienced major delays. Launch was now foreseen in 2013. As for LISA itself, the six spacecraft were cut back to three to save money. A heavy-duty American Atlas rocket would launch all three satellites together sometime in 2018.

On February 3, 2011, the LISA project team presented their plans at a meeting at ESA's headquarters in Paris, France. The hope was that ESA would pick the gravitational wave proposal as the first flagship mission in its new space science program, Cosmic Vision 2015–2025. The other two candidates for this L1 mission (where "L" simply means "large") were also joint ESA/NASA projects: a multi-spacecraft mission to the icy moons of Jupiter, and a large X-ray observatory. On the basis of the Paris presentations and a subsequent recommendation by ESA's Space Science Advisory Committee, the agency would make a final choice in the summer of that same year.

Then disaster struck. On March 15, less than six weeks after the Paris meeting, NASA pulled the plug on all three joint science missions with its European counterpart. The main reasons were the US budget crisis and the soaring costs of the James Webb Space Telescope, the near-infrared successor to Hubble. As for LISA, the project had also failed to receive top priority in the 2010 report *New Worlds, New Horizons,* the National Research Council's assessment of future US projects in the field of astronomy and astrophysics in the decade 2012–2021.

In response to NASA's unfortunate decision, ESA's departing director of science, David Southwood, decided to delay the selection process for the first Cosmic Vision flagship until the spring of 2012. That, he hoped, would give the three teams enough time to come up with a Europe-only alternative, with a much lower price tag. Within a few months, the Einstein wave community drafted a new proposal called NGO, for New Gravitational-Wave Observatory. The arm length was scaled back to 1 million kilometers. As a result, telescopes and mirrors could be smaller, and laser power reduced. The three spacecraft would be less bulky; launch was now possible on two relatively cheap Russian Soyuz boosters. Moreover, NGO sported just one interferometer instead of three. Just as with the original LAGOS design, the "mother spacecraft" would contain the laser and detector equipment (like the central buildings of LIGO and Virgo); the two "daughter spacecraft" just held the end mirrors of the interferometer.

However, when ESA's Science Programme Committee made the final selection for the L1 mission on May 3, 2012, both the NGO proposal and the trimmed-down X-ray proposal were passed over in favor of the Jupiter Icy Moons Explorer. JUICE will be launched in 2022 to study the Jovian satellite Ganymede, an icy world larger than the planet Mercury. It will also make flybys of the satellites Europa and Callisto. Arrival in the Jupiter system is expected in 2030; three years later, the spacecraft will start orbiting Ganymede, the first spacecraft ever to orbit a moon of another planet. Certainly a very exciting mission, but not what gravitational wave scientists had hoped.

LISA Pathfinder project scientist Paul McNamara vividly remembers the gloomy atmosphere during the ninth LISA Symposium, held in Paris in late May 2012. NASA was no longer on board. The LISA International Science Team had been disbanded. Pathfinder was experiencing more delays. Now, even the much smaller, cheaper, and less ambitious NGO mission hadn't been selected for implementation. Everyone was depressed. "It felt like a funeral," says McNamara.

Still, Danzmann and his colleagues didn't give up. After all, the Cosmic Vision program would launch a second flagship mission (L2) in 2028, and there was even a slot for a third (L3) in 2034. A new, independent consortium was formed to keep the scientific community together. In May 2013, the team published a white paper on the expected scientific return of an NGO-like mission with a new name: eLISA, with the *e* standing for *evolved*. "Adding a low-frequency gravitational-wave observatory will add a new sense to our perception of the universe," they wrote in the document's concluding section. "eLISA will be the first ever mission to survey the entire universe with gravitational waves. [It] will play a unique role in the scientific landscape of 2028."

The team's perseverance eventually paid off. In November 2013, ESA announced the science themes for the L2 and L3 missions. L2 would be devoted to high-energy astrophysics (the X-ray mission); L3 was to be reserved for a gravitational wave mission. An official mission call was still many years in the future, but at least the agency had committed itself to this particular field of space science. For the first time ever, there was now absolute certainty that a space-based laser interferometer would become a reality, although it wouldn't be operational until a full sixty years after Rai Weiss and Peter Bender had first discussed the concept. Even the fact that the launch of LISA Pathfinder was delayed again, to late 2015, didn't seem to be too problematic anymore—there was time enough.

At the eleventh Edoardo Amaldi Conference on Gravitational Waves, held in late June 2015 in Gwangju, South Korea, the mood among space scientists was definitely positive. "Reports of LISA's death have been greatly exaggerated," British astrophysicist Jonathan Gair told his audience, paraphrasing a famous Mark Twain quote. Gair went on to describe the fantastic scientific potential of the mission. Simon Barke of Danzmann's Albert Einstein Institute in Hannover was equally optimistic. In one of his presentation slides, he wrote out eLISA as "*evolving* Laser Interferometer Space Antenna," instead of "evolved." Who knows, the mission could become more ambitious again. Or maybe it could launch five years earlier, in 2029, to coincide with the 150th anniversary of Albert Einstein's birth. Barke even suggested that NASA might rejoin the program, especially if LISA Pathfinder was successful, or if the ground-based interferometers bagged their first direct detection. Of course, he could not know that GW150914 was less than three months away.

NASA was not completely out of sight, by the way. The agency had expressed interest in taking part in the eLISA mission, contributing up to $150 million. NASA also took part in the LISA Pathfinder program. US scientists had their own drag-free and attitude control system and their own microthrusters, installed on the same spacecraft and working with the same test masses, but using a different technology. If you fly a demonstrator mission for a giant, expensive future space observatory, it makes sense to test various approaches, of course.

Then came the magical year 2016.

On January 22, seven weeks after its successful launch, LISA Pathfinder arrived at its operational position, some 1.5 million kilometers from the Earth, in the direction of the Sun. Thursday, February 11 saw the triumphant announcement by LIGO and Virgo scientists of the detection of GW150914. Now the whole world knew about gravitational waves. Five days later, Pathfinder's two gold-and-platinum cubes were both released and began floating freely inside their tight enclosures (they had been held still by mechanical clamps and "fingers" during the launch and cruise phases). Regular science operations started on March 1.

Before long, it was evident that LISA Pathfinder was exceeding everyone's expectations. The shielded interior of the spacecraft was indeed the quietest place in the solar system. The net residual acceleration between the two test masses turned out to be about one hundredth of one quadrillionth of the gravitational acceleration here on Earth. In case that doesn't mean much to you, it's the acceleration produced by a force that's equal to the terrestrial weight of an *E. coli* bacterium. That's close enough to an undisturbed free fall to enable future detection of low-frequency Einstein waves. Moreover, Pathfinder's laser interferometer was able to measure the distance between the two cubes to a precision of 35 femtometers (3.5×10^{-14} meter)—much better than required.

These spectacular initial results were published in *Physical Review Letters* on June 7. Less than two weeks later, NASA's L3 Study Team published its interim report online. The team had been established in late 2015 to draw up an inventory of possible technical US contributions to ESA's third flagship mission. One of the team's findings was that a "meaningful participation of the U.S. community in the design, development, and operation of L3 will result in a mission that is more technically robust and more scientifically capable." The interim report suggested ways for NASA to rejoin the program as a junior partner. That fitted in nicely with the conclusions of an earlier report by ESA's Gravitational Observatory Advisory Team (GOAT).

On August 15, there was another important boost, this time from the National Research Council. In their midterm assessment report on the *New Worlds, New Horizons* decadal survey, the panel of authors strongly recommended that NASA restore support to eLISA during this decade and help restore the mission to its original full capacity.

The report's front page featured an artist's impression of gravitational waves and a graphic showing the wave pattern of GW150914. "We made sure the key message was on the cover," said panel member Neil Cornish of Montana State University a few weeks later at a meeting in Zurich, Switzerland.

That meeting was the eleventh LISA Symposium, held close to the Swiss Federal Institute of Technology (formerly the Swiss Federal Polytechnic School), where Albert Einstein studied physics and mathematics at the end of the nineteenth century. If the ninth LISA Symposium had felt like a funeral, this was more like a rebirth party. The atmosphere was energetic, especially when ESA's director of science, Alvaro Giménez, announced that the L3 mission call would be brought forward from 2018 to October 2016. Mission proposals were now expected in January 2017; a final decision could then possibly be made as early as 2020. "We want to make your dreams come true," Giménez told the gravitational wave scientists at the meeting. "Although 2029 is probably too optimistic, we might be able to launch the mission a few years earlier than originally planned, somewhere in the early 2030s."

Paul Hertz, the director of NASA's astrophysics division, offered his full support. Acknowledging that "2011 saw the dissolution of our original LISA partnership," he said, "I'm here to move forward from that." Hertz also said he was quite certain that the mission would get a very strong recommendation in the National Research Council's next report on the decade ahead, in 2020, provided scientists could come up with a convincing mission proposal.

The next day, Wednesday, September 7, the eLISA consortium and NASA's L3 Study Team held their first joint meeting, discussing various options of adapting and improving the original mission plan. NASA's financial contribution was not expected to be restored to 50 percent, but even a few hundred million dollars would make a big difference. There were a lot of possible options: bigger telescopes, more powerful lasers, and longer arm lengths (2 million kilometers, or maybe even 5 million). In Danzmann's words, "We have to come up with a proposal that blows people away." Also, it would be great to go back to three interferometers instead of just one by having lasers in each of the three spacecraft. The two-armed V-shape of eLISA could, everyone hoped, be restored to a full triangle. "We want to have the third arm back," said Danzmann to great enthusiasm, "and we *will* have the third arm back."

MIT physicist David Shoemaker, who has been working in the field since 1975 and who is now leader of Advanced LIGO, was all smiles. "This is a very important meeting," he said. "It really feels like a turning point for eLISA. I would suggest to drop the *e* from now on. There's only one LISA again."

It will still be some time before a final decision is made on the design of the Laser Interferometer Space Antenna (the mission proposal was submitted on January 13, 2017). But one thing's for sure: in the early 2030s, laser beams will bounce between three spacecraft flying in formation around the Sun, most likely a few million kilometers apart, their distances measured to the nearest picometer. Finally, astronomers will be able to detect Einstein waves at millihertz frequencies, monitoring compact binary systems and coalescing supermassive black holes out to the edge of the observable universe.

And LISA may not be alone. At the Zurich meeting, Shuichi Sato of Hosei University in Tokyo, Japan, gave a status report on DECIGO, the Deci-hertz Interferometer Gravitational Wave Observatory. The ambitious plan dates back to 2001. DECIGO would be something like a mini-LISA, with three small satellites separated by a thousand kilometers or so. Somewhere in the next decade, a smaller demonstrator (called pre-DECIGO), with 100-kilometer arms, could be launched into Earth orbit. The full mission would then follow in the 2030s.

Meanwhile, Chinese scientists have plans for *two* space-based interferometers. The first one, called TianQin, has been proposed by a team at Zhongshan University in Guangdong. It would consist of three spacecraft in Earth orbit, forming a giant triangle with the Earth at its center. The interferometer arms would be about 150,000 kilometers

Artist's impression of one of the three spacecraft of the future Laser Interferometer Space Antenna (LISA). Solar panels are on top; laser beams are bounced across millions of kilometers between the individual spacecraft to detect minute path length variations due to passing gravitational waves.

long. A larger Sun-orbiting mission called Taiji is being developed by the Chinese Academy of Sciences. With arms of 3 million kilometers, it's pretty similar to LISA. According to Gang Jin of the academy's Institute of Mechanics, the two Chinese projects may be merged into one mission, to be launched in the early 2030s.

What the space interferometers will actually observe is anybody's guess. Well, not really, of course: astronomers have a reasonable idea about what's going on out there. But details are sketchy. Take coalescing supermassive black holes, for instance. If most galaxies have monster black holes at their centers, and if galaxies crash into their neighbors, you'd expect their black holes to end up orbiting each other in the core of the merged galaxy. At first, they will only produce nano-hertz gravitational waves, detectable in long-term high-precision timing measurements of radio pulsars, as described in Chapter 13. Then, if the two black holes spiral in toward each other, the orbital period decreases and the Einstein wave frequency goes up. A couple of years before they collide and merge, LISA should be able to detect them, no matter how remote they are.

But because light takes time to traverse the universe, galaxies at distances of billions of light-years are observed as they were billions of years ago. So to predict the expected rate of supermassive black hole collisions, astronomers need to know the evolutionary history of both galaxies and their central black holes. They also need to know how likely it is that every binary supermassive black hole will eventually lead to a collision. Theorists have come up with a wide range of predictions, based on a variety of astrophysical assumptions, and no one knows the right answer.

The great thing about LISA, of course, is that the gravitational wave observations will *provide* the answer. Any viable theory about galaxy and black hole evolution needs to be compatible with the observed merger rate. After a few years' worth of operation, LISA will have told us which theories are definitely wrong and which ones may still be right.

The situation is even more uncertain for compact objects that plunge into single supermassive black holes. Every now and then, a supermassive black hole at the core of a galaxy devours a star or a gas cloud that ventures too close. For an average galaxy like ours, this is expected to happen once every few million years. A normal star, one like our own Sun, will almost certainly be ripped apart by the black hole's tidal forces. Some X-ray outbursts that have been observed in other galaxies are probably produced by such tidal disruption events. But a much more compact object, such as a white dwarf star, a neutron star, or a relatively low-mass black hole, is likely to withstand the tidal effects. If it ends up orbiting the supermassive black hole ever faster and faster, the gravitational waves produced by the doomed object can be picked up by LISA. Such an event is called an extreme mass ratio inspiral, or EMRI for short, because the voracious black hole is so much more massive than its lunchtime snack.

The problem is that no one knows how often EMRIs occur. Estimates vary between zero and thousands per year. There are just too many unknowns: the mass distribution of supermassive black holes (how many of them are there within a given mass range), the number of compact objects at the centers of galaxies, the precise dynamics, et cetera. Maybe the compact objects won't end up orbiting the central black hole at all but instead just plunge into oblivion. Again, observations by LISA will provide astronomers with answers. Whatever the observed EMRI rate turns out to be, it will shed valuable light on what goes on—and what doesn't—at the center of galaxies all over the universe.

The same is true for white dwarf binaries in our own galaxy. As you've read in Chapter 5, every Sun-like star ends its life as a white dwarf—an object about as massive as the Sun, but not much larger than the Earth. Since the majority of Milky Way stars are part of binary or multiple systems, you'd expect a very large number of binary white dwarfs, too. If they orbit each other close and fast enough, they will continuously produce Einstein waves in LISA's frequency band. (If they're located in other galaxies, they're probably too far away to generate detectable spacetime ripples.)

Over the past decades, astronomers have discovered a number of such systems. One particularly interesting white dwarf binary is SDSS J065133.338+284423.37, or J0651 for short. It's located at a distance of some 3,500 light-years in the constellation Gemini. The two dwarf stars are separated by a mere 100,000 kilometers or so—about a quarter of the distance between the Earth and the Moon. They orbit each other once every 12.75 minutes, so they're expected to generate gravitational waves with a frequency of 2.6 millihertz, smack in the middle of LISA's sensitivity range. Moreover, astronomers *know* that the system is making waves: the orbital period is decreasing by 0.29 milliseconds per year. In fact, J0651 will serve as a verification source for LISA, like a small number of other compact binaries.

However, no one knows the total number of such tight white dwarf binaries in the Milky Way. LISA is expected to provide a full census, adding enormously to our knowledge about binary star evolution in general and white dwarf properties in particular.

———————

By now, you may wonder how LISA will ever be able to distinguish between all those gravitational wave sources and to derive their individual properties. It's hard enough for LIGO to detect isolated events; how could anyone possibly make sense of dozens or even hundreds of continuous Einstein wave sources, all shaking up spacetime in their own particular way? If GW150914 was a single and clearly distinguishable crack of a whip, then a Milky Way filled with white dwarf binaries is like a ballroom filled with humming tops. Wouldn't you expect LISA's test masses to move chaotically, at many different frequencies at once?

In fact, it's not that bad. Yes, there will be a large number of simultaneous gravitational wave signals superimposed on each other, but it's relatively easy to decompose the apparently chaotic waveform into its constituent sine waves. In fact, your brain is doing the same thing all the time. Your eardrums respond to a large number of different sound waves simultaneously. Still, you have no problem in distinguishing between a human voice, your cellphone ringtone, and a passing car, even if you hear them at the same time. It's just a matter of data analysis.

Some waveforms will of course be harder to recognize, for the simple reason that no one knows what to expect. For instance, cosmologists hope to find evidence for the existence of cosmic strings— weird, one-dimensional structures with a very high mass/energy density that may crisscross our universe. Those topological defects in spacetime are predicted by some theories about the big bang, but nobody knows if they really exist, or what kind of gravitational waves they may produce. Anyway, the LISA data archive is going to be a treasure trove of information for astronomers, high-energy astrophysicists, and cosmologists alike.

To me, one of the most spectacular things about LISA is that it will alert us to upcoming black hole collisions. If LISA had been operational in 2015, astronomers would have known in advance when GW150914 was going to happen, to within a few seconds. Moreover, they would have known exactly where to look for electromagnetic counterparts. All telescopes on the ground and in space would have been monitoring the site of the catastrophe, to see if there might be a simultaneous flash of X-rays, optical emissions, or infrared light. And of course, everybody would have been glued to the TV screens in the LIGO control rooms.

It's not as magical as it may seem. Just before two stellar black holes crash into each other, they have orbital periods of a few milliseconds—that's why they produce high-frequency gravitational waves that can be detected by LIGO and Virgo. But months to years before the cosmic traffic accident, the orbital period is much longer, by seconds or even minutes. Ground-based detectors cannot observe the corresponding low-frequency waves, but LISA can, probably out to billions of light-years.

By studying a continuous-wave source for a long time, the space observatory will be able to triangulate its position in the sky. Large ground-based optical telescopes can then try to identify the host galaxy of the binary, and determine its distance. Meanwhile, a detailed analysis of the waveform provides astronomers with accurate information about the masses of the two objects, and about their orbital evolution. Long before they collide and merge, they have given up most of their secrets. By the time the system's orbital period decreases to just a few seconds, LISA is no longer able to observe it. But by then, it won't be long before a sensitive ground-based interferometer takes over to witness the final stages of the coalescence. You can bet everybody will be on the alert.

It's a taste of things to come. At universities, institutes, and laboratories all around the world, the brightest minds are working hard and passionately to make the Laser Interferometer Space Antenna a reality, ready to be launched in 2031 or so. Some fifteen years from now, LISA—and its Japanese and Chinese equivalents, if they succeed—will revolutionize the field of gravitational-wave astronomy.

Which is not to say that nothing much will happen over the next decade and a half. It's time to focus on more imminent developments. Not in space, but on the ground. Or, rather, below the ground.

Surf's Up for Einstein Wave Astronomy

In a huge cave deep within Mount Ikeno in western Japan, construction workers are building the world's next big laser interferometer. An initial version of KAGRA (Kamioka Gravitational Wave Detector) was completed and tested in March and April 2016. Now, new equipment is being installed to create the baseline version of the instrument. Additional mirrors, towering suspension systems, new lasers, cryogenic cooling units—it's a huge operation, one everyone hopes will be finished by late 2018. If new delays and setbacks can be avoided, that is. Building a 3-kilometer interferometer underground is no mean feat.

Some 200 kilometers farther east, in a suburb of Tokyo, Raffaele Flaminio is optimistic. Yes, there are problems, especially with draining the water that seeps into the caves and tunnels, but those can be solved. They *will* be solved. Flaminio is confident that KAGRA will operate in tandem with LIGO and Virgo as of 2019, maybe even earlier.

Flaminio, an Italian physicist, is director of the Gravitational Wave Project Office of the National Astronomical Observatory of Japan (NAOJ). The good thing about having an Italian director is that the project's meeting room at NAOJ's Mitaka campus serves great coffee.

Flaminio's group is housed in an ugly modern building on a beautiful historic site. Behind the brick gate on Osawa Street, across from a small Buddhist temple, are a few old observatory buildings, surrounded by a nice little Japanese garden, cherry blossoms and all. On weekends, families come here for picnics, unaware of the fact that twenty years ago this was the site of the largest Einstein wave detector in the world.

That was the 300-meter TAMA interferometer, built in 1997, well before LIGO, Virgo, and GEO600. Not only was it the largest prototype detector ever built, it was also the first gravitational wave instrument that was more sensitive than the resonant bar detectors pioneered by Joe Weber and others in the 1960s and 1970s.

Flaminio had been working with the Virgo project since 1990, before the French-Italian project was even approved. He oversaw the construction and commissioning of the interferometer, southeast of Pisa. Between 2004 and 2007, he served as vice director of the European Gravitational Observatory consortium (EGO). By then, he had visited TAMA300 a number of times already, and he had fallen in love with the Land of the Rising Sun. In September 2013, Flaminio moved to Japan.

The Large Cryogenic Gravitational-Wave Telescope (LCGT) was first proposed to the Japanese government soon after the turn of the century. LIGO was about to start its first observations in Hanford and Livingston; Virgo was under construction. Everyone realized that gravitational wave astronomy had a bright future. Japanese scientists wanted to secure their own role in the new field. The idea was to build the detector underground, in the Kamioka mine, to get rid of as much low-frequency seismic noise as possible. Also, the mirrors would be cooled to very low temperatures (hence "cryogenic") to decrease thermal noise. Because of the low temperatures, fused silica is no longer the best choice of material. Instead, the mirrors would be made of ultrapure, synthetically grown sapphire crystals.

After a number of unsuccessful attempts, the project was finally approved in June 2010, shortly after Naoto Kan (the leader of the Democratic Party of Japan) took office as the country's new Prime

Minister. But before construction could really take off, the catastrophic March 11, 2011, Tōhoku earthquake and subsequent mega-tsunami threw a financial monkey wrench into the works. So it wasn't until 2012 that the Kajima Corporation, one of Japan's major construction companies, could start excavating the two 3-kilometer tunnels for what was now known as KAGRA. The job was completed in just two years—according to Flaminio, this was the fastest digging project ever executed in Japan.

Meanwhile, in a nearby cave, the Cryogenic Laser Interferometer Observatory (CLIO) was built—a 100-meter prototype to test the cryogenic mirror technologies. The KAGRA vacuum system was completed in 2015. Most of the interferometer equipment was installed later that year. The first observing run with Initial KAGRA (iKAGRA) took place just a few weeks after LIGO announced the discovery of GW150914, albeit without cryogenics and without the additional mirrors that would create a Fabry-Perot cavity to increase the laser beam's path length and power.

———

From Tokyo's Ueno train station, it's a relaxing two-hour trip on the luxurious Shinkansen bullet train to Toyama on the west coast. From there, an early morning bus takes me into the mountains. It's a stunning seventy-five-minute ride along steep overgrown mountain slopes covered with wisps of fog. I've been told to get off the bus at the post office in Mozumi, a very small mining village just across the border of Gifu prefecture. Mining activities in this region date back to the eighth century. The Kamioka zinc and lead mine—named after another town some ten kilometers down the road—suspended its operations in 2001, but some miner families still live here.

The KAGRA headquarters are located in a new building overlooking the village. Respecting Japanese tradition, I change my shoes for slippers that are provided at the building's entrance. Unfortunately, they're all much too small for my large Dutch feet. Yoichi Aso of NAOJ's Gravitational Wave Project Office shows me around the small control room, where a ten-minute morning briefing takes place with some twenty people present. He then provides me with a hard hat and a reflective safety vest before we drive some five kilometers to the entrance of the mine, about a thousand meters below the summit, in the side of Ikenoyama—Mount Ikeno.

The Kamioka mine has turned into a versatile physics facility over the past decades. In 1991, teams began excavating a huge cave that is now home to Super-Kamiokande, one of the world's largest neutrino detector experiments (that's what the last three letters in the name stand for). It's basically a gigantic stainless steel tank, 41.4 meters high and 39.3 meters in diameter, filled with 50,000 metric tons of ultra-pure water. The cylindrical inner wall of the tank is lined with 11,000 hand-blown photomultiplier tubes, each some 50 centimeters across. They register the faint flashes of light produced by rare interactions of high-energy neutrinos with water molecules. Other underground physics experiments at the Kamioka Observatory are the Kamioka Liquid Scintillator Antineutrino Detector (KamLAND) and the Xenon Detector for Weakly Interacting Massive Particles (XMASS), which is searching for dark matter.

Through yet another horizontal tunnel, Aso takes me to the central area of the KAGRA interferometer, where upgrade activities are in full swing. It's an impressive sight: a huge cavern filled with gleaming vacuum tanks, portal cranes, scaffolding, forklifts, beam pipes with huge bolted flanges, and racks of electronics. Thanks to the stark contrast between the rough-hewn rock and the high-tech equipment, it looks like the secret underground laboratory of some mad scientist in a sci-fi movie. It's very surreal.

The disadvantages of building a gravitational wave laboratory underground are also quite apparent. The rocky walls have been treated with an anti-dust coating, but obviously, the cave can never be as clean as the central buildings of LIGO or Virgo. The most sensitive parts of the equipment are all placed underneath "clean booths": large plastic overpressure tents into which filtered air is blown.

A much bigger problem is water. As any speleologist knows, caves are notoriously wet. The relative humidity in the KAGRA cavern is anywhere between 75 and 100 percent. The mountain acts a lot like a

In July 2016, construction was still very much in progress at the central Laser and Vacuum Equipment Area of the Japanese Kamioka Gravitational-Wave Detector (KAGRA) near the mining town of Mozumi. The walls of the cave have been treated with an anti-dust coating and are covered with plastic to prevent water from dripping onto the equipment.

sponge, explains Aso. It absorbs rainwater, which seeps through the walls of the caves and the two 3-kilometer tunnels. It even wells up through the tunnel floors, thanks to groundwater pressure. It's an incredible amount: some 500 tons of water per hour, on average—1 percent of the volume of the Kamiokande detector.

Aso takes me to the near end of one of the damp, dimly lit tunnels. The stainless steel beam pipes will probably not suffer too much, but water has accumulated into small puddles on the tunnel floor, and I can clearly make out a constant dripping sound. Parts of the ceiling have been covered with huge sheets of plastic. Also, to facilitate drainage, the tunnels are not perfectly horizontal, but slanted by some two degrees. Because of the tilt, the KAGRA mirror surfaces must be slightly tilted, too—another technological challenge.

Most of the walls of the central cavern are also lined with plastic. Here, too, water is the main enemy. The problem was particularly severe in the spring of 2015, when the cave was flooded in some places by up to 10 centimeters of water and the tunnel floors were all wet. Water was dripping from the ceiling of the cave onto the clean booths. The installation of the vacuum system had to be suspended for a couple of months. There had been heavy snows that winter, so there was a lot of meltwater. Maybe the recent excavation of the tunnels with dynamite had also increased groundwater pressure. This year (I'm visiting KAGRA in early July 2016), the situation is less dramatic. Whether this is due to Ikenoyama finding a new equilibrium or to the very strong El Niño climatic event of 2015—which meant much less snow—is anybody's guess. "We'll have to wait and see what happens," says Aso.

Back in Mitaka, Raffaele Flaminio is well aware that the problem is not yet under control. But, he says, the same thing happened in the Gran Sasso underground particle physics laboratory in the Italian Alps. "Just after construction, there was water everywhere. Now it's solved. We will find a solution, too."

When Baseline KAGRA (or bKAGRA) is operational, which is expected to be in late 2018 or early 2019, there will be four large laser interferometers working together. KAGRA is not officially part of the LIGO-Virgo Collaboration. However, in the future, observational data will be shared among the American, European, and Japanese groups for joint analysis. Running four detectors in concert further reduces the false positive rate. Moreover, if the Einstein waves from a neutron star or black hole merger are detected by four independent instruments, pinpointing the event in the sky can be done with relatively high precision. Follow-up observations by automated counterpart searches, as described in Chapter 14, are going to be much more efficient.

Just a few more years in the future, there will be a *fifth* large interferometer, in India. Known as LIGO India, it could be described as an Asian outpost of the LIGO project. Again, the main goal is to improve confidence in future detections by independent confirmation, and to achieve much better localizations. Working toward a global network of detectors has long been a major goal of the Gravitational Wave

International Committee, a panel established in 1997 to increase international collaboration in the field. In early October 2016, a site near the town of Hingoli, some 500 kilometers east of Mumbai, was selected as the future location for the Indian instrument.

Plans for an Indian gravitational wave detector date back to 2009, when physicists established the IndIGO consortium—the Indian Initiative in Gravitational-Wave Observations. Since 2011, discussions have been going on with LIGO officials about relocating American equipment to India. You may remember that the LIGO Hanford Observatory was originally home to two separate interferometers—one with 4-kilometer arms and one with 2-kilometer arms. The same setup had been planned for Advanced LIGO. Obviously, it would be even better to have the second detector at a completely different site, as a third observatory, but that would be much more expensive. On the other hand, the Indian Department of Atomic Energy and the Department of Science and Technology—the main funding agencies for a possible Indian detector—couldn't afford a full-scale project. So why not work together toward LIGO India, with the Indian government paying for the infrastructure and the National Science Foundation for the equipment, roughly speaking?

A similar collaboration had originally been planned between LIGO and a group of physicists from various universities in Australia. However, the Australian government decided to give more priority to the international Square Kilometre Array radio observatory (see Chapter 13); plans for what you might call "LIGO Down Under" never materialized. In the summer of 2012, the National Science Board approved the plan to work with the Indians instead. Indeed, when I visited LIGO Hanford in January 2015, the Laser and Vacuum Equipment Area not only contained the recently completed Advanced LIGO stuff, but also lots of large crates with "spare parts" to be shipped to India as soon as the National Science Foundation (NSF) gave the green light.

An in-principle approval was announced by Prime Minister Narendra Modi on February 17, 2016, just six days after the GW150914

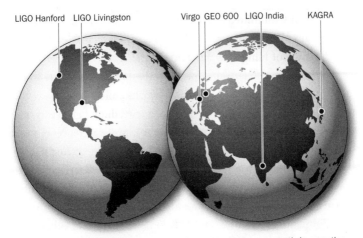

LIGO Hanford LIGO Livingston Virgo GEO 600 LIGO India KAGRA

Locations of the six ground-based laser interferometers that are currently in operation (LIGO Hanford, LIGO Livingston, Virgo, and GEO600), under construction (KAGRA), or planned (LIGO India).

press conference. Six weeks later, on March 31, NSF director France Córdova signed a memorandum of understanding with her Indian colleagues, and LIGO India could proceed. Eventually it will be an almost exact copy of the current Advanced LIGO detectors, with 4-kilometer arms. The hope is that LIGO India will be operational in 2024.

Prediction is very difficult, especially if it's about the future, said Danish scientist Niels Bohr, who was a contemporary of Albert Einstein. In the 1920s, the two great physicists were debating the nature of reality, both in person and in a lengthy exchange of letters. Bohr was a pioneer of quantum physics; Einstein had grave doubts about the consequences of this theory. Neither of them could ever have foreseen that just a century later astronomers would be operating a worldwide network of gravitational wave detectors to learn about extreme events in the universe. And studying Einstein waves from black hole collisions might finally reveal why Einstein's theory of general relativity is still fundamentally incompatible with quantum field theory.

Even now, it's hard to predict what the status of gravitational wave astronomy will be in the mid-2020s. By then, five giant detectors will be on the alert for minute ripples in spacetime. These may be as small as a sextillionth of a percent (one part in 10^{23}) and last anywhere from a fraction of a second up to a minute. Collisions and mergers of neutron stars and black holes, out to distances of a few billion light-years, may well be detected every week or so on average. Small differences between the arrival times of the signal at the five individual detectors will enable a precise triangulation of its direction of origin. Rapid follow-up observations by counterpart search programs yield additional information about the collision event and its host galaxy. Meanwhile, pulsar timing observations by the Square Kilometre Array and other radio observatories will reveal a background of very-low-frequency Einstein waves generated by orbiting supermassive black holes throughout the universe. Many of those nanohertz waves can be traced back to monstrous binaries in relatively nearby galaxies. And polarization measurements of the cosmic microwave background may finally find the "fingerprints" of primordial gravitational waves produced in the very first split second after the big bang.

These are all "expected deliverables" of Einstein wave astronomy. But almost every scientist I have interviewed for this book stressed that the *unexpected* results will probably be the most revolutionary and spectacular. That's the great thing about explorative science: you don't know in advance what you will discover. Past experiences have shown that opening up a new field of research always leads to great surprises. There's no reason to expect that gravitational wave astronomy is going to be the first exception to that rule.

Astronomy is sometimes described as the oldest science. After all, our distant ancestors were already watching the stars and keeping track of the motions of the Sun, the Moon, and the planets. But sometimes I feel that astronomy has only just started. For many thousands of years, our knowledge about the universe was fully dependent on what we could perceive with the naked eye. Only in the past four centuries, after Hans Lipperhey's invention of the telescope, did astronomy really start to flourish. And in the course of those four hundred years, the study of the stars has been an accelerating succession of revolutionary insights, made possible by new discoveries and technological breakthroughs, culminating in the advent of space flight and the digital revolution.

A major recurring theme has been the opening up of new, unexplored parts of the electromagnetic spectrum, from the discovery of infrared light by William Herschel in 1800 to the launch of space telescopes that register the highest-energy gamma rays. Today, we're no longer constrained by the limited sensitivity of the human eye, or by the absorbing effects of the Earth's atmosphere. For the first time ever, we can enjoy the cosmic vista in all its splendor and variety.

I like to compare traditional, pre-telescopic astronomy to being locked up in a thick-walled brick building in one of the most spectacular landscapes our planet has to offer. There's just one small, narrow slit in the eastern wall of the building that offers us a very limited view of the outside world. The only thing we can make out is what looks like a grassy plain in the foreground, a distant hillslope with a couple of trees, and a white cloud in the blue sky. Enough to get a rough idea of what's out there, but way incomplete of course. That's what naked-eye, optical-wavelength astronomy is.

Opening up the other parts of the electromagnetic spectrum is like creating additional breaches in the walls. Not just narrow slits, but large windows. Suddenly, we become aware of the impressive waterfall in the south, and the range of active volcanoes in the west. We see rivers, snow-capped mountains, and rolling thunderclouds. Our limited "slit view" is still an integral part of this impressive scenery of course, but for the first time we learn how it fits in, and we start to discover the underlying geological patterns. Finally, everything starts to make sense.

Infrared astronomy lets us peer deep inside clouds of gas and dust to unravel the birth of stars and planets. Ultraviolet astronomy reveals the extremely tenuous gas in the "empty" space between clusters of galaxies, and teaches us about the physics of the hottest stars in the Milky Way. Millimeter-wave astronomy made us aware of the faint afterglow of the big bang and provides us with insight into the origin of galaxies and the formation of planets. Through radio astronomy, we are able to map out neutral hydrogen atoms—the most common cosmic constituent—throughout the universe. Moreover, radio astronomy introduced us to exotic objects such as pulsars and quasars. Finally, X-ray and gamma-ray astronomy have opened up a window on the hot and violent universe of exploding stars, colliding galaxies, shock waves, and black holes.

Over and over again, exploring a new field has brought unexpected discoveries and revolutionary insights. And in the case of gravitational wave astronomy, there's only *more* reason to expect surprises. That's because we're not just broadening our cosmic vision; we're adding a completely new sense to our means of studying the universe.

In his inspiring lectures, both for fellow scientists and for general audiences and schoolkids, gravitational wave physicist Bernard Schutz (director of the Albert Einstein Institute in Potsdam, Germany, between 1995 and 2014 and now at Cardiff University in Wales) compares today's astronomy to a deaf man's walk through the jungle. Looking around, the man sees trees, ferns, vines, insects, birds, snakes, and monkeys. After a while, if he's an attentive observer, he will learn a lot about his environment. He may even be fooled into believing he knows almost everything there is to know.

But then, by some magic trick, a fairy gives him back his sense of hearing. Suddenly, he is overwhelmed by new information. This is not about seeing details that had been hidden before; it's a completely new sense. The sounds of the jungle—birdsong, rustling leaves, snapping twigs, et cetera—provide much additional information about the things he was already able to see. But being able to hear also provides knowledge about things that are out of sight, like the thunderous crash of a falling tree a kilometer away or the roaring of predatory animals in the distance.

In Schutz's words: "Our universe is a jungle, filled with wild animals. Thanks to gravitational waves, we are beginning to listen to them for the first time." Gravitational wave astronomy is often described as a way of "listening" to the universe. Of course, Einstein waves have nothing to do with sound, but it's a strong and valuable metaphor nonetheless. The real promise of the new field is indeed the discovery of objects and events that cannot be observed at all by studying electromagnetic radiation. Gravitational waves are new cosmic messengers, with new stories to tell.

The hope is that studying the tiny vibrations of spacetime will also add in solving some uncanny mysteries about the universe we live in. For instance, astronomers have found indirect evidence for the existence of huge amounts of dark matter. It's stuff we can't see—dark matter isn't even supposed to consist of normal atoms and molecules—but we can detect its gravity. The outer parts of galaxies rotate much faster than you would expect on the basis of the visible matter in them. The same is true for the velocities of galaxies in clusters. Also, the amount of gravitational lensing by galaxy clusters (the bending of light from background sources by the cluster's gravity) can be explained only if there's a lot of dark matter around. The problem is, no one has a clue as to the nature of dark matter, and despite heroic efforts by particle physicists and cosmologists alike, no direct trace of dark matter has ever been found so far.

A second major mystery is dark energy. Studies of the expansion history of the universe have revealed that space has been growing at an accelerating pace for some 5 billion years. Common sense tells you that the expansion should slow down as a result of the mutual gravity between galaxies. Instead, it's speeding up. The only possible explanation that physicists have been able to come up with is the existence of a mysterious "repulsive" energy in empty space. The concept is not entirely new: it sort of fits in with quantum theory, and Albert Einstein himself introduced a dark-energy-like "cosmological constant" in his equations before Edwin Hubble discovered the expansion of the universe. But again, the true nature of dark energy is anybody's guess.

The gravity of the problem becomes apparent when you realize that dark matter and dark energy together make up 96 percent of the total mass / energy density in the universe. In other words, we only know about a meager 4 percent of what's out there; the rest is an outright riddle. And there doesn't seem to be an easy way out, either. Detailed studies of the cosmic microwave background and of the present large-scale structure of the universe all point to the same conclusion: we only can make sense of our universe if its evolution has been governed by the mysterious forces of dark matter and dark energy.

Future advances in gravitational wave astronomy may offer further exciting clues, especially where dark energy is concerned. The amplitude of gravitational waves from colliding compact objects can be precisely predicted by general relativity. From the observed waveform (the so-called chirp), it's straightforward to calculate the masses of the two coalescing bodies. General relativity then tells you the original amplitude of the Einstein waves that are produced. By comparing that value to the much smaller amplitude measured by detectors here on Earth, it's easy to find the merger's distance.

If counterpart searches identify the merger's host galaxy, optical telescopes can determine the galaxy's redshift. As we saw in Chapter 9, the redshift of a galaxy is a measure of the time it has taken the galaxy's light to reach Earth. It will then become possible to combine redshift measurements and independent distance estimates for a large number of galaxies at various distances. This will reveal the expansion history of the universe: any deceleration or acceleration will produce deviations from a nice linear relationship between distance and redshift. And gaining detailed knowledge about the cosmic expansion history is an important way to learn more about dark energy.

In fact, the first indications of the existence of dark energy, in 1998, were found in a more or less similar way. Astronomers studied a particular kind of supernova explosion (so-called Type Ia supernovae), for which the true energy output is known. Such objects are referred to as "standard candles." Measuring the apparent brightness of the supernova gives information about the distance, which can then be compared to the redshift of the host galaxy. One potential problem with this method is that the apparent brightness of a remote stellar explosion can also be influenced by other effects, such as absorption by dust. In the case of gravitational waves, however, what you see is what you get. The universe is completely transparent to the spacetime ripples, so the observed amplitude can readily be translated into a true distance. If

Type Ia supernovae are standard candles, gravitational waves can be described as "standard sirens."

As for solving the mystery of dark matter, the role of Einstein waves is less evident. But future observations of gravitational waves from merging supermassive black holes or from compact objects plunging into black holes (the so-called extreme mass ratio inspirals, or EMRIs) may help in mapping the clustering of galaxies at various epochs in the history of the universe. Combined with a better knowledge of the expansion history, this could yield more detailed information about the spatial distribution of dark matter, and possibly about the mysterious stuff's true nature.

Finally, physicists look forward to new ways of putting Einstein's theory of general relativity on trial. The study of gravitational waves tells them about the behavior of matter and spacetime under extreme circumstances: the incredibly strong gravitational fields in the immediate surroundings of black holes. Observations of EMRIs in particular are expected to contain a lot of useful information on so-called strong-field environments. As I've mentioned before, the theory of general relativity is incompatible with quantum field theory, so everybody expects that at least one of the two theories is going to falter at some point. They can't both be completely right. The big question is, when and where do the first cracks in either theory appear, and how will physicists be able to repair them? Future gravitational wave observations may lead the way, by putting the squeeze on general relativity.

Some theorists even suspect that all the problems described above are somehow related to each other. Adherents of the theory of modified Newtonian dynamics (MOND) believe that dark matter may be largely an illusion, caused by our misconceptions about gravity. Others expect that a true theory of quantum gravity will automatically solve the riddle of dark energy and the accelerated expansion of the universe. And almost everyone is confident that the long-sought marriage of general relativity and quantum theory will help us understand such enigmatic concepts as black holes, the big bang, and the multiverse. Studying Einstein waves at all possible frequencies and from every corner of the cosmos—listening to the jungle, so to speak—is an important next step in our quest to understand the fundamental properties of the universe. The first direct detection of gravitational waves on September 14, 2015, marked the start of a completely new chapter in the history of astronomy.

———————

Building a huge laser interferometer in space, as described in Chapter 15, will be a very important new development in gravitational wave astronomy. But not all progress is going to occur beyond Earth's atmosphere. The Laser Interferometer Space Antenna will only be sensitive to a certain range of relatively low-frequency ripples, basically determined by its tremendous scale, with arm lengths on the order of a few million kilometers. To keep monitoring the high-frequency waves from the final stages of neutron star and black hole mergers, smaller instruments will always be necessary. And fifteen or twenty years from now, LIGO and Virgo—and perhaps KAGRA—likely will have been superseded by a new generation of ground-based detectors.

Even before the upgrade to Advanced Virgo began, European scientists came up with the first ideas for a third-generation interferometer. It's called the Einstein Telescope, or ET for short. Like KAGRA, the Einstein Telescope will have cryogenically cooled mirrors. It will have the same triangular layout as LISA, but with arm lengths of 10 kilometers. It will contain a total of six interferometers, with lasers, beam splitters, mirrors, and light detectors at each vertex. Three of the six interferometers (one per vertex) will be sensitive to gravitational waves with frequencies between 2 and 40 hertz; the other three will focus on higher-frequency waves.

Because Europe is so densely populated, it's hard to find a good location for such a large detector, so the plan is to build it in caves and tunnels underground. The added bonus is the lower sensitivity of an underground detector to low-frequency seismic noise. Because of its longer arms, lower noise levels, and cryogenic mirrors, ET is expected to be dozens of times more sensitive than Advanced Virgo. As a result, it will be able to detect neutron star and black hole mergers throughout all of the observable universe, out to distances of more than 13 billion light-years.

In 2010 and 2011, a preliminary design study for the ambitious project was carried out, with funds from the Seventh Framework program of the European Commission. ET is already one of the "Magnificent Seven" European projects recommended by the ASPERA network for the future development of astroparticle physics in Europe. It could be operational in the early 2030s, around the time of the launch of LISA.

Of course, Europe is not alone in thinking about a next-generation instrument. In 2013, at the Gravitational Waves Advanced Detectors Workshop on the Italian island of Elba, a small group of American scientists came up with a plan for an even larger ground-based instrument: the Long Ultra-Low-Noise Gravitational-Wave Observatory (LUNGO). According to Matt Evans of the Massachusetts Institute of Technology, it all started half jokingly, during a late-night conversation. Hastily they put together a PowerPoint presentation for one of the next day's sessions.

Since then, the proposal has gained quite a bit of momentum. It now goes by the name of Cosmic Explorer. Another appropriate name would be Super-LIGO, since it will have the same L-shaped design as its predecessor, albeit with arm lengths of 40 kilometers instead of 4. Outfitted with cryogenically cooled mirrors, Cosmic Explorer would be even more sensitive than the Einstein Telescope. And since the United States has so many wide-open spaces left, there's no immediate need to go underground. According to Evans, the salt flats east of Carson City, Nevada (well known for the car and motorcycle races held there), would be a great location for the future interferometer. But there are many other possibilities—the western states contain a lot of federally owned land. "During another late-night conversation at the 2016 Advanced Detectors Workshop, we even considered the possibility of building Cosmic Explorer at sea," says Evans. "Maybe it's not crazy."

Scaling up Virgo and LIGO by a factor of three or even ten will obviously increase the required budget, too. A ballpark cost estimate for both the Einstein Telescope and the Cosmic Explorer is around $1 billion, comparable to the price tag of other large science facilities, such as the existing Atacama Large Millimeter / submillimeter Array (ALMA), the emerging Square Kilometre Array radio observatory (SKA), and the future European Extremely Large Telescope (E-ELT). Large-scale international collaboration is probably necessary to realize such ambitious goals. Ideally, there would be a worldwide network of third-generation instruments, maybe with a large triangular detector in Europe, a huge L-shaped interferometer in the United States, and another big L in the Southern Hemisphere. Maybe the last of these could be built in Western Australia after all—the Australian International Gravitational Research Centre (AIGRC) in Perth is still a very active member of the LIGO Scientific Collaboration, even though plans for a smaller Australian LIGO were put on the backburner earlier this century.

What about the location for the European Einstein Telescope? That won't probably be decided for another couple of years. However, some scientists involved in the project do have their preferences. Physicist Jo van den Brand of the Dutch National Institute for Subatomic Physics (Nikhef) in Amsterdam was born in the far southeast of the country, close to the tripoint between the Netherlands, Belgium, and Germany. The area is well known for its coal mining activities in the twentieth century. Van den Brand believes it's an ideal location to build ET. Seismic test measurements have already revealed that the subsurface rock is very stable. Meanwhile, the loess upper layer, consisting of windblown silt particles, is a very good insulator against surface vibrations.

Sites in Hungary, Spain, and on the Italian island of Sardinia are also being considered, but the Albert Einstein Institute in Hannover is supporting the choice for the Dutch-German border region. As a Dutchman, I find it hard not to get tremendously excited about the possibility, however remote it may be, that my home country will host the Einstein Telescope. We'll know soon enough.

The human quest to understand the universe is a never-ending endeavor. That's the great thing about science: every answer raises new questions, and the search for more and deeper knowledge will never be completed. The hunt for gravitational waves is a textbook example of scientific exploration, spanning a full hundred years from the initial theoretical prediction to the first direct detection. It has been a volatile adventure of self-confident pioneers and persevering scientists, dreams and nightmares, setbacks and successes, technological challenges, and unwavering passion and drive.

Albert Einstein once said: "Look deep into nature, and then you will understand everything better." The same is true of what gravitational wave astronomy is giving us. We've learned to surf the waves of spacetime. The journey is far from over—it's only just beginning.

Notes and Further Reading

1. A SPACETIME APPETIZER

6   The movie *Interstellar,* directed by Christopher Nolan and starring
    Matthew McConaughey, Anne Hathaway, Jessica Chastain, and
    Michael Caine, was distributed in North America by Paramount
    Pictures and released in the United States on November 5, 2014.

8–10   A great overview of the history of astronomy is Timothy Ferris,
       Coming of Age in the Milky Way (New York: William Morrow & Co.,
       1988).

10–12   A good general and up-to-date introduction to the universe is Neil
        deGrasse Tyson, Michael A. Strauss, and J. Richard Gott, *Welcome to
        the Universe: An Astrophysical Tour* (Princeton, NJ: Princeton Univer-
        sity Press, 2016).

16   See Albert Einstein, *Relativity: The Special and the General Theory*
     (London: Methuan & Co., 1920, originally published in German in
     1916); George Gamow, *Mr. Tompkins in Wonderland* (Cambridge:
     Cambridge University Press, 1940); Eva Fenyo, *A Guided Tour
     through Space and Time* (Upper Saddle River, NJ: Prentice-Hall,
     1959), and Kip S. Thorne, *Black Holes and Time Warps: Einstein's
     Outrageous Legacy* (New York: W.W. Norton & Co., 1994). A classic
     and humorous introduction to higher dimensions is Edwin Abbott
     Abbott, *Flatland: A Romance of Many Dimensions* (London: Seeley &
     Co., 1884).

20   Kip Thorne, *The Science of Interstellar* (New York: W.W. Norton &
     Co., 2014). See also Oliver James, Eugenie von Tunzelmann, Paul
     Franklin, and Kip S. Thorne, "Visualizing Interstellar's Wormhole,"

American Journal of Physics 83, no. 486 (2016) (doi: 10.1119/1.4916949) and, by the same authors, "Gravitational Lensing by Spinning Black Holes in Astrophysics, and in the Movie *Interstellar,*" *Classical and Quantum Gravity* 32, no. 6 (2015) (doi: 10.1088/0264-9381/32/6/065001).

2. RELATIVELY SPEAKING

22   More about the Wall Poems of Leiden can be found here: http://www.muurgedichten.nl/wallpoems.html. The website for Museum Boerhaave, Leiden, the Netherlands, is http://www.museumboerhaave.nl/english.

23   I visited the depot of Museum Boerhaave in Leiden on April 7, 2016.

26   The movie of Apollo 15 astronaut David Scott dropping a feather and a hammer on the Moon can be found here: https://www.youtube.com/watch?v=KDp1tiUsZw8

29   The discovery of Neptune is described in Tom Standage, *The Neptune File* (London: Penguin Books, 2000).

30–31   Urbain Le Verrier's hunt for an intramercurial planet is described in Thomas Levenson, *The Hunt for Vulcan . . . And How Albert Einstein Destroyed a Planet, Discovered Relativity, and Deciphered the Universe* (New York: Random House, 2015).

38   The collected papers of Albert Einstein can be found here: http://einsteinpapers.press.princeton.edu

There are numerous biographies about Albert Einstein. One of the most comprehensive is by Abraham Pais, *Subtle Is the Lord: The Science and Life of Albert Einstein* (Oxford: Oxford University Press, 1982). Another great Einstein biography is Dennis Overbye, *Einstein in Love: A Scientific Romance* (New York: Viking Penguin, 2000). See also Brian Greene, *The Fabric of the Cosmos: Space, Time, and the Texture of Reality* (New York: Alfred A. Knopf, 2004).

3. EINSTEIN ON TRIAL

40   I interviewed Francis Everitt at Stanford University, California, on June 20, 2016.

40–41,
54–58   Gravity Probe B: http://einstein.stanford.edu

44–48   More on Arthur Eddington's eclipse expedition of 1919 can be found in Peter Coles, "Einstein, Eddington and the 1919 Eclipse," *Proceedings of International School on "The Historical Development of Modern Cosmology," Valencia 2000,* ASP Conference Series (https://arxiv.org/abs/astro-ph/0102462).

48   Gaia mission: http://sci.esa.int/gaia

More on testing general relativity can be found in Amanda Gefter, "Putting Einstein to the Test," *Sky & Telescope* 110, no. 1 (July 2005): 33. See also Clifford M. Will, *Was Einstein Right? Putting Relativity to the Test* (New York: Basic Books, 1986; 2nd edition, 1993), and, by the same author, "Was Einstein Right? Testing, Relativity at the Centenary," *Annals of Physics* 15, no. 1–2 (January 2006): 19–33 (doi: 10.1002/andp.200510170).

4. WAVE TALK AND BAR FIGHTS

72   I interviewed Tony Tyson at the University of California at Davis on June 20, 2016.

73   Dick Garwin is portrayed in Joel Shurkin, *True Genius: The Life and Work of Richard Garwin* (New York: Penguin Random House, 2017).

Joe Weber's early work on the detection of gravitational waves is described in Marcia Bartusiak, *Einstein's Unfinished Symphony: The Story of a Gamble, Two Black Holes, and a New Age of Astronomy* (New Haven: Yale University Press, 2017), and in Janna Levin, *Black Hole Blues and Other Songs from Outer Space* (New York: Alfred A. Knopf, 2016). A thorough introduction to the history of gravitational wave physics is Daniel Kennefick, *Traveling at the Speed of Thought: Einstein and the Quest for Gravitational Waves* (Princeton, NJ: Princeton University Press, 2007). A very detailed account of the origins of gravitational wave research, including Joe Weber's experiments, is Harry Collins, *Gravity's Shadow: The Search for Gravitational Waves* (Chicago: University of Chicago Press, 2004).

5. THE LIVES OF STARS

77   *Cosmos: A Personal Voyage* is a thirteen-part TV series written by Carl Sagan, Ann Druyan, and Steven Soter and directed by Adrian Malone. It was first broadcast by PBS between September 28 and December 21, 1981. The title of this chapter is also the title of the ninth episode. See the book as well: Carl Sagan, *Cosmos* (New York: Random House, 1980).

A good introduction to stellar evolution is James B. Kaler, *Cosmic Clouds: Birth, Death, and Recycling in the Galaxy* (New York: W.H. Freeman & Co., 1996); see also Kaler, *Stars and Their Spectra: An Introduction to the Spectral Sequence* (Cambridge: Cambridge University Press, 1989; 2nd edition, 2011), and Kaler,

Heaven's Touch: From Killer Stars to the Seeds of Life, How We Are Connected to the Universe (Princeton, NJ: Princeton University Press, 2009). A detailed but accessible introduction to neutron stars is Werner Becker, ed., *Neutron Stars and Pulsars* (New York: Springer, 2009).

6. CLOCKWORK PRECISION

95–97   Jocelyn Bell's own story on the discovery of pulsars can be read here: http://www.bigear.org/vol1no1/burnell.htm

99   Arecibo Observatory: http://www.naic.edu

105–106   I interviewed Joel Weisberg by telephone on August 2, 2016.

106   Nobel Prize in Physics 1993: https://www.nobelprize.org/nobel_prizes/physics/laureates/1993; Nobel Prize in Physics 1974: https://www.nobelprize.org/nobel_prizes/physics/laureates/1974

107   Marta Burgay et al., "An Increased Estimate of the Merger Rate of Double Neutron Stars from Observations of a Highly Relativistic System," *Nature* 426 (4 December 2003): 531–533 (doi: 10.1038/nature02124).

109–110   Freeman Dyson's prediction of the production of gravitational waves by coalescing neutron stars was published in A. G. W. Cameron, ed., *Interstellar Communication. A Collection of Reprints and Original Contributions* (New York: W.A. Benjamin, 1963).

111   Joel M. Weisberg, Joseph H. Taylor, and Lee A. Fowler, "Gravitational Waves from an Orbiting Pulsar," *Scientific American* 245, no. 4 (October 1981): 74–82 (doi: 10.1038/scientificamerican1081-74).

A good introduction to pulsars is Geoff McNamara, *Clocks in the Sky: The Story of Pulsars* (New York: Springer, 2008). See also Duncan R. Lorimer, "Binary and Millisecond Pulsars," *Living Reviews in Relativity,* 8 (2005): 7 (doi: 10.12942/lrr-2005-7).

7. LASER QUEST

112   My visit to the LIGO Livingston Observatory in Louisiana in the spring of 1998 was funded by the Dutch weekly magazine *Intermediair.*

112–113   I visited the LIGO Hanford Observatory (Washington) and interviewed Frederick Raab on January 14, 2015.

116–126   More on laser interferometry: https://www.ligo.caltech.edu/page /ligo-gw-interferometer. A nice movie by Marco Kraan of the Dutch National Institute for Subatomic Physics (Nikhef) on laser interferometry can be found here: https://www.youtube.com/watch ?v=h_FbHipV3No

8. THE PATH TO PERFECTION

129   I interviewed Rai Weiss in Seattle on January 6, 2015 and by telephone on June 29, 2016.

129–143   Interview with Rai Weiss by Shirley Cohen, Caltech Oral History Project: http://oralhistories.library.caltech.edu/183/1/Weiss_OHO.pdf

132   Rainer Weiss, "Electronically Coupled Broadband Gravitational Antenna," *Quarterly Progress Report,* Research Laboratory of Electronics (MIT), no. 105 (1972): 54 (http://www.hep.vanderbilt.edu /BTeV/test-DocDB/0009/000949/001/Weiss_1972.pdf)

133   Cosmic Background Explorer (COBE): http://science.nasa.gov /missions/cobe. See also Charles W. Misner, Kip S. Thorne, and John Archibald Wheeler, *Gravitation* (New York: W.H. Freeman & Co., 1973).

136   Paul Linsay, Peter Saulson, Rainer Weiss, and Stan Whitcomb, *A Study of a Long Baseline Gravitational Wave Antenna System* (the LIGO Blue Book) (National Science Foundation: 1983) (https://dcc.ligo.org/public/0028/T830001/000/NSF_bluebook _1983.pdf).

138   Rochus E. Vogt, Ronald W. P. Drever, Kip S. Thorne, Frederick J. Raab, and Rainer Weiss, *A Laser Interferometer Gravitational-Wave Observatory: Proposal to the National Science Foundation* (California Institute of Technology, December 1989) (https://dcc.ligo.org /public/0065/M890001/003/M890001-03%20edited.pdf).

140   I interviewed Tony Tyson at the University of California at Davison June 20, 2016.

142   I interviewed Barry Barish at the California Institute of Technology in Pasadena on June 22, 2016.

142–143   Interview with Barry Barish by Shirley Cohen, Caltech Oral History Project: http://oralhistories.library.caltech.edu/178/1 /Barish_OHO.pdf

143–146   I visited the Virgo detector at the European Gravitational Observatory in Santo Stefano a Macerata, near Pisa, Italy, and interviewed Federico Ferrini on September 22, 2015.

146–147   I visited the Albert Einstein Institute in Hannover, Germany, and interviewed Karsten Danzmann and Bruce Allen on August 4 and 5, 2016.

147–148   I visited the GEO600 detector in Ruthe, near Hannover, Germany, on February 9, 2015.

Website of LIGO: https://www.ligo.caltech.edu. Website of the LIGO Scientific Collaboration: http://ligo.org. Website of Virgo: http://public.virgo-gw.eu/language/en. Website of the European Gravitational Observatory (EGO): http://www.ego-gw.it. Website of GEO600: http://www.geo600.org. The history of LIGO is described in Marcia Bartusiak, *Einstein's Unfinished Symphony: The Story of a Gamble, Two Black Holes, and a New Age of Astronomy* (New Haven: Yale University Press, 2017). See also Harry Collins, *Gravity's Shadow: The Search for Gravitational Waves* (Chicago: University of Chicago Press, 2004), and Janna Levin, *Black Hole Blues and Other Songs from Outer Space* (New York: Alfred A. Knopf, 2016).

9. CREATION STORIES

See the following books for more information: Joseph Silk, *The Big Bang* (New York: W.H. Freeman & Co., 1980; 3rd edition, 2001); Simon Singh, *Big Bang: The Most Important Scientific Discovery of All Time and Why You Need to Know about It* (New York: Fourth Estate, 2004); George Smoot and Keay Davidson, *Wrinkles in Time: The Imprint of Creation* (London: Little, Brown and Company, 1993), and Dennis Overbye, *Lonely Hearts of the Cosmos: The Story of the Scientific Quest for the Secret of the Universe* (New York: HarperCollins, 1991).

10. COLD CASE

167–171   My visit to McMurdo Station and the Amundsen-Scott South Pole Station on Antarctica in December 2012 was organized and funded by the National Science Foundation, as part of their Antarctic Journalist Program.

168   E and B Experiment (EBEX): http://groups.physics.umn.edu/cosmology/ebex

169   IceCube: https://icecube.wisc.edu. South Pole Telescope: https://pole.uchicago.edu

169–171   Background Imaging of Cosmic Extragalactic Polarization (BICEP): http://bicepkeck.org

171   Cosmic Background Explorer (COBE): http://science.nasa.gov/missions/cobe

172 Wilkinson Microwave Anisotropy Probe (WMAP): http://science
.nasa.gov/missions/wmap. Planck: http://sci.esa.int/planck. Nobel
Prize in Physics 2006: https://www.nobelprize.org/nobel_prizes
/physics/laureates/2006

172–173 I visited the Llano de Chajnantor and the Atacama Large Millimeter / submillimeter Array (ALMA) (http://www.almaobservatory
.org) in northern Chile in 1998 (facilitated by the National
Radio Astronomy Observatory, NRAO), 1999 (sponsored by the
European Southern Observatory, ESO), 2004, 2007 (funded by
ESO and the Dutch Research School for Astronomy, NOVA),
2010, 2012 (sponsored by ESO), 2013 (sponsored by ESO), and
in 2015 and 2017 (both as a tour guide for the Dutch monthly
magazine *New Scientist*).

173 Atacama Cosmology Telescope (ACT): http://act.princeton.edu

178–180 BICEP2 press conference at the Harvard-Smithsonian Center
for Astrophysics on March 17, 2014: https://www.youtube.com
/watch?v=Iasqtm1prlI

178–184 I interviewed John Kovac by telephone on June 30, 2016; I interviewed Christine Pulliam by telephone on July 1, 2016.

181 The video of Chao-Lin Kuo bringing the BICEP2 news to Andrei
Linde and his wife Renata Kallosh is here: https://www.youtube
.com/watch?v=ZlfIVEy_YOA

183 BICEP2 science meeting at the Harvard-Smithsonian Center
for Astrophysics on March 17, 2014: https://www.youtube.com
/watch?v=0n9NPvEbJr0. P. A. R. Ade et al. (BICEP2 / Keck and
Planck Collaborations), "Joint Analysis of BICEP2 / Keck Array
and Planck Data," *Physical Review Letters* 114 (2015): 101301
(doi: 10.1103/PhysRevLett.114.101301).

For more information, see Alan H. Guth, *The Inflationary Universe: The Quest
for a New Theory of Cosmic Origins* (New York: Basic Books, 1998).

11. GOTCHA

187–190 I interviewed Marco Drago by telephone on July 11, 2016.

190–191,
197–198 I interviewed Stan Whitcomb at the California Institute of Technology in Pasadena on June 23, 2016.

191–194,
203 I interviewed Gabriela González in Zürich, Switzerland, on

September 7, 2016; I interviewed David Reitze in Amsterdam, the Netherlands, on March 2, 2016.

193–194,
201   I interviewed Lawrence Krauss by telephone on July 29, 2016.

194–196   The two major blind injections in the LIGO and Virgo data are described in much detail in Harry Collins, *Gravity's Ghost and Big Dog: Scientific Discovery and Social Analysis in the Twenty-First Century* (Chicago: University of Chicago Press, 2011). More on blind injections can be found here: http://www.ligo.org/news /blind-injection.php

199–204   I interviewed Whitney Clavin at the California Institute of Technology in Pasadena on June 22, 2016.

200   More on the discovery of GW150914: *LIGO Magazine* 8 (March 2016) (http://www.ligo.org/magazine/LIGO-magazine -issue-8.pdf). B. P. Abbott et al. (LIGO Scientific Collaboration and Virgo Collaboration), "Observation of Gravitational Waves from a Binary Black Hole Merger," *Physical Review Letters* 116 (2016): 061102 (doi: 10.1103/PhysRevLett.116.061102).

202   Joshua Sokol, "Latest Rumour of Gravitational Waves Is Probably True This Time," *New Scientist,* February 8, 2016 (https://www .newscientist.com/article/2076754-latest-rumour-of-gravitational -waves-is-probably-true-this-time).

202–203   GW150914 press conference at the National Press Club, Washington, DC: https://www.youtube.com/watch?v=aEPIwEJmZyE

202–204   I interviewed France Córdova by telephone on June 28, 2016.

205   Special Breakthrough Prize in Fundamental Physics 2016: https://breakthroughprize.org/News/32. 2016 Gruber Foundation Cosmology Prize: http://gruber.yale.edu/cosmology/press/2016 -gruber-cosmology-prize-press-release. Shaw Prize in Astronomy 2016: http://www.shawprize.org/en/shaw.php?tmp=3&twoid =102&threeid=254&fourid=476. 2016 Kavli Prize in Astrophysics: http://www.kavliprize.org/prizes-and-laureates/prizes /2016-kavli-prize-astrophysics. 2016 Amaldi Medal: http://public .virgo-gw.eu/adalberto-giazotto-guido-pizzella-share-amaldi -medal

Two recent books that also cover the first direct detection of gravitational waves are Harry Collins, *Gravity's Kiss: The Detection of Gravitational Waves* (Cambridge, MA: MIT Press, 2017), and Marcia Bartusiak, *Einstein's Unfinished Symphony: The Story of a Gamble, Two Black Holes, and a New Age of Astronomy* (New Haven: Yale University Press, 2017).

12. BLACK MAGIC

209   The movie of the merging black holes that produced GW150914 is here: https://www.youtube.com/watch?v=Zt8Z_uzG71o

210–212   B. P. Abbott et al. (LIGO Scientific Collaboration and Virgo Collaboration), "Astrophysical Implications of the Binary Black Hole Merger GW150914," *Astrophysical Journal Letters* 818 (2016): L22 (http://iopscience.iop.org/article/10.3847/2041-8205/818/2/L22). See also B. P. Abbott et al. (LIGO Scientific Collaboration and Virgo Collaboration), "Properties of the Binary Black Hole Merger GW150914," *Physical Review Letters* 116 (2016): 241102 (doi: 10.1103/PhysRevLett.116.241102).

212   Gabriela González, Fulvio Ricci, and David Reitze, "Latest News from the LIGO Scientific Collaboration," press conference from American Astronomical Society (AAS) 228th meeting, 15 June 2016, San Diego, California: https://aas.org/media-press/archived-aas-press-conference-webcasts. See also B. P. Abbott et al. (LIGO Scientific Collaboration and Virgo Collaboration), "GW151226: Observation of Gravitational Waves from a 22-Solar-Mass Binary Black Hole Coalescence," *Physical Review Letters* 116 (2016): 241103 (doi: 10.1103/PhysRevLett.116.241103).

224   I interviewed Gijs Nelemans at Radboud University in Nijmegen, the Netherlands, on July 14, 2016.

On black holes, see Kip S. Thorne, *Black Holes and Time Warps: Einstein's Outrageous Legacy* (New York: W.W. Norton & Co., 1994); Igor Novikov, *Black Holes and the Universe* (Cambridge: Cambridge University Press, 1995), and Clifford A. Pickover, *Black Holes: A Traveler's Guide* (New York: John Wiley & Sons, Inc., 1996). See also this interactive website on black holes: http://hubblesite.org/explore_astronomy/black_holes.

13. NANOSCIENCE

226–227   Parkes Observatory: https://www.parkes.atnf.csiro.au

229   Don C. Backer, Shrinivas R. Kulkarni, Carl Heiles, M. M. Davis and W. Miller Goss, "A Millisecond Pulsar," *Nature* 300 (December 16, 1982): 615–618 (http://www.nature.com/nature/journal/v300/n5893/abs/300615a0.html).

230   Aleksander Wolszczan and Dale A. Frail, "A Planetary System around the Millisecond Pulsar PSR1257+12," *Nature* 355 (January 9, 1992): 145–147 (http://www.nature.com/nature/journal/v355/n6356/abs/355145a0.html).

237–238   Parkes Pulsar Timing Array (PPTA): http://www.atnf.csiro.au /research/pulsar/ppta

238   European Pulsar Timing Array (EPTA): http://www.epta.eu.org. Westerbork Synthesis Radio Telescope: http://www.astron.nl/radio -observatory/public/public-0. North American Nanohertz Observatory for Gravitational Waves (NANOGrav): http://nanograv.org

239   International Pulsar Timing Array (IPTA): http://www.ipta4gw .org; Stephen R. Taylor et al., "Are We There Yet? Time to Detection of Nanohertz Gravitational Waves Based on Pulsar-Timing Array Limits," *Astrophysical Journal Letters* 819, no. 1 (2016): L6 (doi: 10.3847/2041-8205/819/1/L6).

241   Large European Array for Pulsars (LEAP): http://www.epta.eu.org /leap.html

242–245   Square Kilometre Array: https://www.skatelescope.org

243   Australian SKA Pathfinder (ASKAP): http://www.atnf.csiro.au /projects/askap/index.html. Murchison Widefield Array (MWA): http://www.mwatelescope.org/

243–244   I visited the Murchison Radio Observatory in Western Australia on November 28, 2012 together with Marieke Baan of the Dutch Research School for Astronomy (NOVA), and again on June 15, 2016 during a trip that was funded by the Australian Department of Foreign Affairs and Trade.

244   Hydrogen Epoch of Reionisation Array radio telescope (HERA): https://www.ska.ac.za/science-engineering/hera. MeerKAT array: https://www.ska.ac.za/science-engineering/meerkat. I visited the HERA telescope and the MeerKAT array in South Africa on November 24 and 25, 2016. See also Sarah Wild, *Searching African Skies: The Square Kilometre Array and South Africa's Quest to Hear the Songs of the Stars* (Sunnyside, South Africa: Jacana Media, 2012).

14. FOLLOW-UP QUESTIONS

246   I visited the Roque de los Muchachos Observatory on La Palma, Spain on a number of occasions between 1996 and 2016.

247–248   Govert Schilling, *Flash! The Hunt for the Biggest Explosions in the Universe* (Cambridge: Cambridge University Press, 2002). Jonathan I. Katz, *The Biggest Bangs: The Mystery of Gamma-Ray Bursts, the Most Violent Explosions in the Universe* (Oxford: Oxford University Press, 2002).

249   I visited ESO's Paranal Observatory in northern Chile in 1998, 1999 (sponsored by the European Southern Observatory, ESO), 2004, 2007 (funded by ESO and the Dutch Research School for Astronomy, NOVA), 2010 (as a tour guide for SNP Natuurreizen), 2012 (sponsored by ESO), and 2015 and 2017 (both as a tour guide for the Dutch monthly magazine *New Scientist*).

253   MeerLICHT telescope: http://www.ast.uct.ac.za/meerlicht

253, 264   I interviewed Paul Groot at Radboud University in Nijmegen, the Netherlands, on July 14, 2016.

258   Nial R. Tanvir et al., "A 'Kilonova' Associated with the Short-Duration γ-Ray Burst GRB 130603B," *Nature* 500 (August 3, 2013): 547 (https://arxiv.org/abs/1306.4971).

261–262   Swift gamma-ray burst mission: http://swift.gsfc.nasa.gov

262   Fermi gamma-ray space telescope: http://fermi.gsfc.nasa.gov. Panoramic Survey Telescope & Rapid Response System (PanSTARRS): https://www.ifa.hawaii.edu/research/Pan-STARRS.shtml. Dark Energy Camera: http://www.ctio.noao.edu/noao/node/1033

262–263   I visited the Palomar Observatory in California on June 21, 2016.

263   I interviewed Eric Bellm at the California Institute of Technology in Pasadena on June 22, 2016.

263–264   Zwicky Transient Facility: http://www.ptf.caltech.edu/ztf

264–265   BlackGEM: https://astro.ru.nl/blackgem. I visited the La Silla Observatory in northern Chile in 1987, 2004, 2010 (as a tour guide for SNP Natuurreizen), 2012 (sponsored by ESO) and 2013.

266   Large Synoptic Survey Telescope (LSST): https://www.lsst.org. Evryscope: http://evryscope.astro.unc.edu

15. SPACE INVADERS

267–269   My trip to Kourou, French Guiana, to attend the launch of LISA Pathfinder on December 3, 2015, was organized and funded by the European Space Agency (ESA).

267–272   LISA Pathfinder: http://sci.esa.int/lisa-pathfinder

268   I visited the LISA Pathfinder clean room at the Industrieanlagen-Betriebsgesellschaft in Ottobrunn, Germany, on September 1, 2015. I interviewed Paul McNamara in Zürich, Switzerland, on September 6, 2016.

274   Space Antenna for Gravitational-Wave Astronomy (SAGA): http://adsabs.harvard.edu/full/1985ESASP.226..157F

276   National Research Council, *New Worlds, New Horizons in Astronomy and Astrophysics* (Washington, DC: National Academies Press, 2010) (https://www.nap.edu/catalog/12951/new-worlds-new-horizons-in-astronomy-and-astrophysics).

276–277   Cosmic Vision 2015–2025: http://sci.esa.int/cosmic-vision. New Gravitational-Wave Observatory (NGO): http://sci.esa.int/ngo

277   Jupiter Icy Moons Explorer (JUICE): http://sci.esa.int/juice. Pau Amaro-Seoane et al., *Doing Science with eLISA: Astrophysics and Cosmology in the Millihertz Regime,* eLISA White Paper (https://www.elisascience.org/dl/1201.3621v1.pdf). Evolved Laser Interferometer Space Antenna (eLISA): https://www.elisascience.org

278   I attended the eleventh Edoardo Amaldi Conference on Gravitational Waves in Gwangju, South Korea, from June 22–26, 2015.

279   M. Armano et al., "Sub-Femto-g Free Fall for Space-Based Gravitational Wave Observatories: LISA Pathfinder Results," *Physical Review Letters* 116 (June 7, 2016): 231101. L3 Study Team Interim Report: https://pcos.gsfc.nasa.gov/studies/L3/L3ST_Interim_Report-Final.pdf. Gravitational Observatory Advisory Team, *The ESA L3 Gravitational-Wave Mission,* Final Report (http://sci.esa.int/cosmic-vision/57910-goat-final-report-on-the-esa-l3-gravitational-wave-mission).

279–280   National Research Council, *New Worlds, New Horizons: A Midterm Assessment,* (Washington, DC: National Academies Press, 2016) (https://www.nap.edu/catalog/23560/new-worlds-new-horizons-a-midterm-assessment).

280–281   I attended the eleventh LISA Symposium in Zürich, Switzerland, on September 6 and 7, 2016.

281   LISA Mission Proposal (January 2017): https://www.elisascience.org/files/publications/LISA_L3_20170120.pdf. Deci-hertz Interferometer Gravitational-Wave Observatory (DECIGO): http://tamago.mtk.nao.ac.jp/decigo/index_E.html

16. SURF'S UP FOR EINSTEIN WAVE ASTRONOMY

288–289   I visited the Mitaka campus of the National Astronomical Observatory of Japan (NAOJ) and interviewed Raffaele Flaminio in Tokyo, Japan, on July 6, 2016.

289   TAMA300: http://tamago.mtk.nao.ac.jp/spacetime/tama300_e.html

290–293   I visited the Kamioka Gravitational-Wave Detector (KAGRA) near Mozumi, Gifu prefecture, Japan, on July 7, 2016.

291–293   Kamioka Gravitational-Wave Detector (KAGRA): http://gwcenter .icrr.u-tokyo.ac.jp/en

293–295   LIGO India: http://www.gw-indigo.org/ligo-india

294   Indian Initiative in Gravitational-Wave Observations (IndIGO): http://www.gw-indigo.org/tiki-index.php

298   I attended a public talk by Bernard Schutz in Gwangju, South Korea, on June 25, 2015.

302–304   Einstein Telescope: http://www.et-gw.eu

303   I interviewed Matt Evans by telephone on June 30, 2016.

303–304   B. P. Abbott et al., *Exploring the Sensitivity of Next Generation Gravitational Wave Detectors,* LIGO Document LIGO-P1600143 (https://arxiv.org/pdf/1607.08697v3.pdf).

For more on dark matter and dark energy, see Robert P. Kirshner, *The Extravagant Universe: Exploding Stars, Dark Energy and the Accelerating Cosmos* (Princeton: Princeton University Press, 2002); Iain Nicolson, *Dark Side of the Universe: Dark Matter, Dark Energy, and the Fate of the Cosmos* (Bristol: Canopus Publishing Ltd., 2007), and Richard Panek, *The 4 Percent Universe: Dark Matter, Dark Energy, and the Race to Discover the Rest of Reality* (Boston: Houghton Mifflin Harcourt, 2011).

Acknowledgments

I am grateful to the following scientists for giving me the opportunity to interview them as part of the research for this book (either in person or by telephone): Bruce Allen, Barry Barish, Eric Bellm, Joan Centrella, Whitney Clavin, Harry Collins, France Córdova, Karsten Danzmann, Marco Drago, Anamaria Effler, Matt Evans, Francis Everitt, Raffaele Flaminio, Peter Fritschel, Neil Gehrels, Joe Giaime, Gabriela González, Paul Groot, Vincent Icke, Gemma Janssen, Mansi Kasliwal, John Kovac, Lawrence Krauss, Avi Loeb, Jess McIver, Maura McLaughlin, Paul McNamara, Gijs Nelemans, Stirl Phinney, Tsvi Piran, Christine Pulliam, Frederick Raab, Christian Reichardt, David Reitze, Jean-Paul Richard, David Shoemaker, Ira Thorpe, Virginia Trimble, Tony Tyson, Jo van den Brand, Chris Van den Broeck, Jeroen van Dongen, Alan Weinstein, Joel Weisberg, Rainer Weiss, and Stan Whitcomb. I am also grateful for helpful comments on the drafts of a number of chapters from Dirk van Delft and Jeroen van Dongen (Chapter 2), Joel Weisberg (Chapter 6), Joris van Heijningen (Chapter 7), David Shoemaker (Chapters 7 and 8), Gabriela González (Chapter 11), Gijs Nelemans (Chapter 12), Gemma Janssen (Chapter 13), and Paul McNamara (Chapter 15). The thoughtful comments of a small number of anonymous reviewers have helped me to improve the manuscript even further. Finally, I am indebted to Martin Rees for writing the book's foreword.

Illustration Credits

Index

REYNOLDS PRICE

THREE GOSPELS

THE GOOD NEWS ACCORDING TO MARK

THE GOOD NEWS ACCORDING TO JOHN

AN HONEST ACCOUNT OF A MEMORABLE LIFE

SCRIBNER

NEW YORK   LONDON   TORONTO   SYDNEY   TOKYO   SINGAPORE

S C R I B N E R
1230 Avenue of the Americas
New York, NY 10020

Portions of "An Honest Account of a Memorable Life"
previously appeared in Theology Today;
portions of the preface to "The Good News According to John"
appeared in Incarnation: Contemporary Writers
on the New Testament (Viking, 1990).

Set in Adobe Electra

Manufactured in the United States of America

5   7   9   10   8   6   4

Library of Congress Cataloging-in-Publication Data
Bible. N.T. Mark. English. Price. 1996.
The three gospels / Reynolds Price.
p.   cm.
Contents: The good news according to Mark—The good news
according to John—An honest account of a memorable life.
1. Jesus Christ—Fiction. 2. Bible N.T. Gospels—Paraphrases, English.
I. Price, Reynolds, date. II. Bible. N.T. John. English. Price. 1996. IV. Title.
BS2583.P74   1996
226.3'05209—dc20   95-39948
CIP
ISBN 0-684-80336-4

FOR

ERIK BENSON